SpringerBriefs in Electrical and Computer Engineering

Series editors

Woon-Seng Gan, School of Electrical and Electronic Engineering, Nanyang Technological University, Singapore, Singapore

C.-C. Jay Kuo, University of Southern California, Los Angeles, CA, USA

Thomas Fang Zheng, Research Institute of Information Technology, Tsinghua University, Beijing, China

Mauro Barni, Department of Information Engineering and Mathematics, University of Siena, Siena, Italy

T0171956

SpringerBriefs present concise summaries of cutting-edge research and practical applications across a wide spectrum of fields. Featuring compact volumes of 50 to 125 pages, the series covers a range of content from professional to academic. Typical topics might include: timely report of state-of-the art analytical techniques, a bridge between new research results, as published in journal articles, and a contextual literature review, a snapshot of a hot or emerging topic, an in-depth case study or clinical example and a presentation of core concepts that students must understand in order to make independent contributions.

More information about this series at http://www.springer.com/series/10059

F. Richard Yu • Ying He

Deep Reinforcement Learning for Wireless Networks

 Springer

F. Richard Yu
Carleton University
Ottawa, ON, Canada

Ying He
Carleton University
Ottawa, ON, Canada

ISSN 2191-8112   ISSN 2191-8120   (electronic)
SpringerBriefs in Electrical and Computer Engineering
ISBN 978-3-030-10545-7   ISBN 978-3-030-10546-4   (eBook)
https://doi.org/10.1007/978-3-030-10546-4

Library of Congress Control Number: 2018967964

This Springer imprint is published by the registered company Springer Nature Switzerland AG.
The registered company address is: Gewerbestrasse 11, 6330 Cham, Switzerland

Preface

A Brief Journey Through "Deep Reinforcement Learning for Wireless Networks"

There is a phenomenal burst of research activities in machine learning and wireless systems. Machine learning evolved from a collection of powerful techniques in AI areas and has been extensively used in data mining, which allows the system to learn the useful structural patterns and models from training data. Reinforcement learning is an important branch of machine learning, where an agent learns to take actions that would yield the most reward by interacting with the environment. The main advantage of reinforcement learning is that it works well without prior knowledge of an exact mathematical model of the environment. However, the traditional reinforcement learning approach has some shortcomings, such as low convergence rate to the optimal behavior policy and its inability to solve problems with high-dimensional state space and action space. These shortcomings can be addressed by deep reinforcement learning. The key idea of deep reinforcement learning is to approximate the value function by leveraging the powerful function approximation property of deep neural networks. After training the deep neural networks, given a state-action pair as input, deep reinforcement learning is able to estimate the long-term reward. The estimation result can guide the agent to choose the best action. Deep reinforcement learning has been successfully used to solve many practical problems. For example, Google DeepMind adopts this method on several artificial intelligent projects with big data (e.g., AlphaGo) and gets quite good results. In this book, we utilize deep reinforcement learning approach to wireless networks to improve the system performance.

As the first chapter of this book, Chap. 1, we introduce machine learning algorithms, which are classified into four categories: supervised, unsupervised, semi-supervised, and reinforcement learning. In this chapter, widely used machine learning algorithms are introduced. Each algorithm is briefly explained with some examples.

Chapter 2 focuses on reinforcement learning and deep reinforcement learning. A brief review of reinforcement learning and Q-learning is first described. Then, recent advances of deep Q-network are presented, and double deep Q-network and dueling deep Q-network that go beyond deep Q-network are also given.

Chapter 3 presents a deep reinforcement learning approach in cache-enabled opportunistic interference alignment wireless networks. Most existing works on cache-enabled interference alignment wireless networks assume that the channel is invariant, which is unrealistic considering the time-varying nature of practical wireless environments. In this chapter, we consider realistic time-varying channels. The complexity of the system is very high when we consider realistic time-varying channels. We use Google TensorFlow to implement deep reinforcement learning in this chapter to obtain the optimal interference alignment user selection policy in cache-enabled opportunistic interference alignment wireless networks. Simulation results are presented to show that the performance of cache-enabled opportunistic interference alignment networks in terms of the network's sum rate and energy efficiency can be significantly improved by using the proposed approach.

Chapter 4 considers trust-based mobile social networks with mobile edge computing, in-network caching, and device-to-device communications. An optimization problem is formulated to maximize the network operator's utility with comprehensive considerations of trust values, computation capabilities, wireless channel qualities, and the cache status of all the available nodes. We apply a deep reinforcement learning approach to automatically make a decision for optimally allocating the network resources. The decision is made purely through observing the network's states, rather than any handcrafted or explicit control rules, which makes it adaptive to variable network conditions. Simulation results with different network parameters are presented to show the effectiveness of the proposed scheme.

The book will be useful to both researchers and practitioners in this area. The readers will find the rich set of references in each chapter particularly valuable.

Ottawa, ON, Canada   F. Richard Yu
October 2018   Ying He

Contents

Chapter 1
Introduction to Machine Learning

Abstract Machine learning is evolved from a collection of powerful techniques in AI areas and has been extensively used in data mining, which allows the system to learn the useful structural patterns and models from training data. Machine learning algorithms can be basically classified into four categories: supervised, unsupervised, semi-supervised and reinforcement learning. In this chapter, widely-used machine learning algorithms are introduced. Each algorithm is briefly explained with some examples.

A machine learning approach usually consists of two main phases: training phase and decision making phase as illustrated in the Fig. 1.1. At the training phase, machine learning methods are applied to learn the system model using the training dataset. At the decision making phase, the system can obtain the estimated output for each new input by using the trained model. Figure 1.2 shows supervised, unsupervised, semi-supervised and reinforcement learning, which are described in the following.

1.1 Supervised Learning

Supervised learning is a kind of labelling learning technique. Supervised learning algorithms are given a labeled training dataset (i.e., inputs and known outputs) to build the system model representing the learned relation between the input and output. After training, when a new input is fed into the system, the trained model can be used to get the expected output [1, 2]. In the following, we will give a detailed representation of widely-used supervised learning algorithms, such as k-nearest neighbor, decision tree, random forest, neural network, support vector machine, Bayes' theory, and hidden markov models.

F. R. Yu, Y. He, *Deep Reinforcement Learning for Wireless Networks*, SpringerBriefs in Electrical and Computer Engineering, https://doi.org/10.1007/978-3-030-10546-4_1

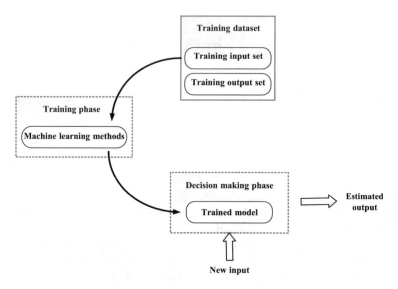

Fig. 1.1 The general processing procedure of a machine learning approach

1.1.1   k-Nearest Neighbor (k-NN)

The k-NN is a supervised learning technique, where the classification of a data sample is determined based on the k nearest neighbors of that unclassified sample. The process of the k-NN algorithm is very simple: if the most of the k nearest neighbors belong to a certain class, the unclassified sample will be classified into that class. Figure 1.3 shows a simple example of how k-NN algorithm works. Specially, when $k = 1$, it becomes the nearest neighbor algorithm. The higher the value of k is, the less effect the noise will have on the classification. Since the distance is the main metric of the k-NN algorithm, several functions can be applied to define the distance between the unlabeled sample and its neighbors, such as Chebyshev, City-block, Euclidean and Euclidean squared. For a more insightful discussion on k-NN, please refer to [3].

1.1.2   Decision Tree (DT)

The DT is one of the classification techniques which performs classification through a learning tree. In the tree, each node represents a feature (attribute) of a data, all branches represent the conjunctions of features that lead to classifications, and each leaf node is a class label. The unlabeled sample can be classified by comparing its feature values with the nodes of the decision tree [4, 5]. The DT has many advantages, such as intuitive knowledge expression, simple implementation and high classification accuracy. ID3 [6], C4.5 [7] and CART [8] are three widely-used decision tree algorithms to perform the classification of training dataset

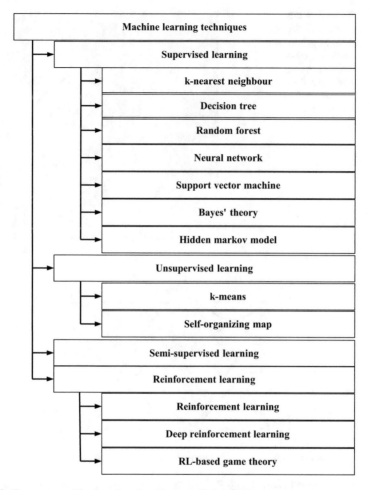

Fig. 1.2   Common machine learning algorithms applied to SDN

automatically. The biggest difference among them is the splitting criteria which are used to build decision trees. The splitting criteria used by ID3, C4.5 and CART are Information gain, Gain ratio and Gini impurity respectively. Reference [9] gives a detailed comparison of the three DT algorithms.

1.1.3   Random Forest

The random forest method [10], also known as random decision forest, can be used for classification and regression tasks. A random forest consists of many decision trees. To mitigate over-fitting of decision tree method and improve accuracy, the

Fig. 1.3 Example of k-NN algorithm, for $k = 5$. Among the five closest neighbors, one neighbor belongs to class A and four neighbors belong to class B. In this case, the unlabeled example will be classified into class B

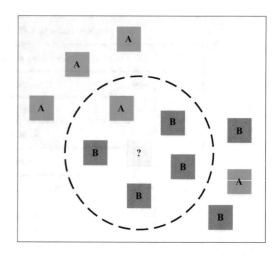

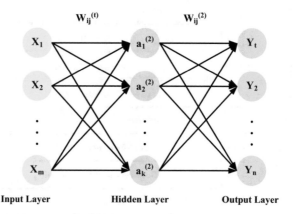

Input Layer   **Hidden Layer**   **Output Layer**

Fig. 1.4 A basic neural network with three layers: an input layer, a hidden layer and an output layer. An input has m features (i.e., $X_1, X_2, \ldots, X_m$) and the input can be assigned to n possible classes (i.e., $Y_1, Y_2, \ldots, Y_n$). Also, W_{ij}^l denotes the variable link weight between the ith neuron of layer l and the jth neuron of layer $l + 1$, and a_k^l denotes the activation function of the kth neuron in layer l

random forest method randomly chooses only a subset of the feature space to construct each decision tree. The steps to classify a new data sample by using random forest method are: (a) put the data sample to each tree in the forest. (b) Each tree gives a classification result, which is the tree's "vote". (c) The data sample will be classified into the class which has the most votes.

1.1.4 Neural Network (NN)

A neural network is a computing system made up of a large number of simple processing units, which operate in parallel to learn experiential knowledge from historical data [11]. The concept of neural networks is inspired by the human brain, which uses basic components, known as neurons to perform highly complex, nonlinear and parallel computations. In a NN, its nodes are the equivalent components of the neurons in the human brain. These nodes use activation functions to perform nonlinear computations. The most frequently used activation functions are the sigmoid and the hyperbolic tangent functions [12]. Simulating the way neurons are connected in the human brain, the nodes in a NN are connected to each other by variable link weights.

A NN has many layers. The first layer is the input layer and the last layer is the output layer. Layers between the input layer and the output layer are hidden layers. The output of each layer is the input of the next layer and the output of the last layer is the result. By changing the number of hidden layers and the number of nodes in each layer, complex models can be trained to improve the performance of NNs. NNs are widely used in many applications, such as pattern recognition. The most basic NN has three layers, including an input layer, a hidden layer and an output layer, which is shown in Fig. 1.4.

There are many types of neural networks, which are often divided into two training types, supervised or unsupervised [13]. In the following, we will give a detailed representation of supervised neural networks which have been applied in the field of SDN. In Sect. 1.2.2, self-organizing map, a representative type of unsupervised neural networks, will be described.

Random NN

The random NN can be represented as an interconnected network of neurons which exchange spiking signals. The main difference between random NN and other neural networks is that neurons in random NN exchange excitatory and inhibitory spiking signals probabilistically. In random NN, the internal excitatory state of each neuron is represented by an integer, which is called "potential". The potential value of each neuron rises when it receives an excitatory spiking signal and drops when it receives an inhibitory spiking signal. Neurons whose potential values are strictly positive are allowed to send out excitatory or inhibitory spiking signals to other neurons according to specific neuron-dependent spiking rates. When a neuron sends out a spiking signal, its potential value drops one. The random NN has been used in classification and pattern recognition [14]. For a more insightful discussion on random NN, please refer to [14–16].

Deep NN

Neural networks with a single hidden layer are generally referred to as shallow NNs. In contrast, neural networks with multiple hidden layers between the input layer and the output layer are called deep NNs [17–19]. For a long time, shallow NNs are often used. To process high-dimensional data and to learn increasingly complex models, deep NNs with more hidden layers and neurons are needed. However, deep NNs increase the training difficulties and require more computing resources. In recent years, the development of hardware data processing capabilities (e.g., GPU and TPU) and the evolved activation functions (e.g., ReLU) make it possible to train deep NNs [20]. In deep NNs, each layer's neurons train a feature representation based on the previous layer's output, which is known as feature hierarchy. The feature hierarchy makes deep NNs capable of handling large high-dimensional datasets. Due to the multiple-level feature representation learning, compared to other machine learning techniques, deep NNs generally provide much better performance [20].

Convolutional NN

Convolutional NN and recurrent NN are two major types of deep NNs. Convolutional NN [21, 22] is a feed-forward neural network. Local sparse connections among successive layers, weight sharing and pooling are three basic ideas of convolutional NN. Weight sharing means that weight parameters of all neurons in the same convolution kernel are same. Local sparse connections and weight sharing can reduce the number of training parameters. Pooling can be used to reduce the feature size while maintaining the invariance of features. The three basic ideas reduce the training difficulties of convolutional NNs greatly.

Recurrent NN

In feed-forward neural networks, the information is transmitted directionally from the input layer to the output layer. However, recurrent NN [23, 24] is a stateful network, which can use internal state (memory) to handle sequential data. A typical recurrent NN and its unrolled form are shown in Fig. 1.5. X_t is the input at time step t. h_t is the hidden state at time step t. h_t captures information about what happened in all the previous time steps, so it is called "memory". Y_t is the output at time step t. U, V and W are parameters in the recurrent NN. Unlike a traditional deep NN, which uses different parameters at each layer, the recurrent NN shares the same parameters (i.e., U, V and W) across all time steps. This means that at each time step, the recurrent NN performs the same task, just with different inputs. In this way, the total number of parameters needed to be trained is reduced greatly. Long Short-Term Memory (LSTM) [25, 26] is the most commonly-used type of recurrent NNs, which has a good ability to capture long-term dependencies. LSTM uses three gates (i.e., an input gate, an output gate and a forget gate) to compute the hidden state.

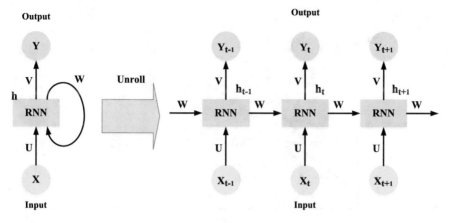

Fig. 1.5 A typical recurrent NN and its unrolled form. X_t is the input at time step t. h_t is the hidden state at time step t. Y_t is the output at time step t. U, V and W are parameters in the recurrent NN

1.1.5 Support Vector Machine (SVM)

SVM is another popular supervised learning method, invented by Vapnik and others [27], which has been widely used in classification and pattern recognition. The basic idea of SVM is to map the input vectors into a high-dimensional feature space. This mapping is achieved by applying different kernel functions, such as linear, polynomial and Radial Based Function (RBF). Kernel function selection is an important task in SVM, which has effect on the classification accuracy. The selection of kernel function depends on the training dataset. The linear kernel function works well if the dataset is linearly separable. If the dataset is not linearly separable, polynomial and RBF are two commonly-used kernel functions. In general, the RBF-based SVM classifier has a relatively better performance than the other two kernel functions [28, 29].

The objective of SVM is to find a separating hyperplane in the feature space to maximize the margin between different classes. Note that, the margin is the distance between the hyperplane and the closest data points of each class. The corresponding closest data points are defined as support vectors. An example of SVM classifier is shown in Fig. 1.6. From the figure, there are many possible separating hyperplanes between two classes, but only one optimal separating hyperplane can maximize the margin. For a more insightful discussion on SVM, please refer to [30–32].

1.1.6 Bayes' Theory

Bayes' theory uses the conditional probability to calculate the probability of an event occurring given the prior knowledge of conditions that might be related to the event. The Bayes' theory is defined mathematically as the following equation:

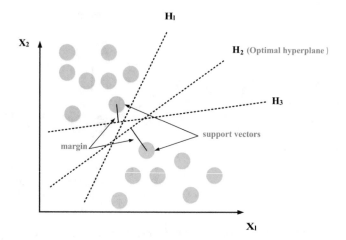

Fig. 1.6 An example of SVM classifier with an optimal linear hyperplane. There are two classes in the figure, and each class has one support vector. As it can be seen, there are many possible separating hyperplanes between two classes, such as H_1, H_2 and H_3, but only one optimal separating hyperplane (i.e., H_2) can maximize the margin

$$P(H|E) = \frac{P(E|H)P(H)}{P(E)} \tag{1.1}$$

where E is a new evidence, H is a hypothesis, $P(H|E)$ is the posterior probability that the hypothesis H holds given the new evidence E, $P(E|H)$ is the posterior probability that of evidence E conditioned on the hypothesis H, $P(H)$ is the prior probability of hypothesis H, independent of evidence E, and $P(E)$ is the probability of evidence E.

In a classification problem, the Bayes' theory learns a probability model by using the training dataset. The evidence E is a data sample, and the hypothesis H is the class to assign for the data sample. The posterior probability $P(H|E)$ represents the probability of a data sample belonging to a class. In order to calculate the posterior probability $P(H|E)$, $P(H)$, $P(E)$ and $P(E|H)$ need to be calculated first based on the training dataset using the probability and statistics theories, which is the learning process of the probability model. When classifying a new input data sample, the probability model can be used to calculate multiple posterior probabilities for different classes. The data sample will be classified into the class with the highest posterior probability $P(H|E)$. The advantage of the Bayes' theory is that it requires a relatively small number of training dataset to learn the probability model [33]. However, there is an important independence assumption when using the Bayes' theory. To facilitate the calculation of $P(E|H)$, the features of data samples in the training dataset are assumed to be independent of each other [34]. For a more insightful discussion on Bayes' theory, please refer to [33, 35–38].

1.1.7 Hidden Markov Models (HMM)

HMM is one kind of Markov models. Markov models are widely used in randomly dynamic environments which obey the memoryless property. The memoryless property of Markov models means that the conditional probability distribution of future states only relates to the value of the current state and is independent of all previous states [39, 40]. There are other Markov models, such as Markov Chains (MC). The main difference between HMM and other models is that HMM is often applied in environments where system states are partially visible or not visible at all.

1.2 Unsupervised Learning

In contrast to supervised learning, an unsupervised learning algorithm is given a set of inputs without labels (i.e., there is no output). Basically, an unsupervised learning algorithm aims to find patterns, structures, or knowledge in unlabeled data by clustering sample data into different groups according to the similarity between them. The unsupervised learning techniques are widely used in clustering and data aggregation [2, 41]. In the following, we will give a detailed representation of widely-used unsupervised learning algorithms, such as k-means and self-organizing map.

1.2.1 k-Means

The k-means algorithm is a popular unsupervised learning algorithm, which is used to recognize a set of unlabeled data into different clusters. To implement the k-means algorithm, only two parameters (i.e., the initial dataset and the desired number of clusters) are needed. If the desired number of clusters is k, the steps to resolve node clustering problem by using k-means algorithm are: (a) initialize k cluster centroids by randomly choosing k nodes; (b) use a distance function to label each node with the closest centroid; (c) assign new centroids according to the current node memberships and (d) stop the algorithm if the convergence condition is valid, otherwise go back to step (b). An example procedure of k-means algorithm is shown in Fig. 1.7. For a more insightful discussion on k-means, please refer to [2, 42].

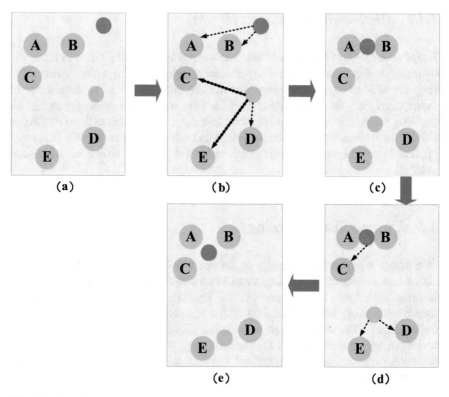

Fig. 1.7 Example of k-means algorithm, for $k = 2$. (**a**) Randomly choosing two data points as two centroids; (**b**) label each node with the closest centroid, resulting that node A and B are a class, node C, D and E are another class; (**c**) assign new centroids; (**d**) label each node with the closest centroid again, resulting that node A, B and C are a class, node D and E are another class; (**e**) the algorithm is converged

1.2.2 Self-Organizing Map (SOM)

SOM, also known as Self-Organizing Feature Map (SOFM), is one of the most popular unsupervised neural network models. SOM is often applied to perform dimensionality reduction and data clustering. In general, SOM has two layers, an input layer and a map layer. When SOM is used to perform data clustering, the number of neurons in the map layer is equal to the desired number of clusters. Each neuron has a weight vector. The steps to resolve data clustering problem by using SOM algorithm are: (a) initialize the weight vector of each neuron in the map layer; (b) choose a data sample from the training dataset; (c) use a distance function to calculate the similarity between the input data sample and all weight vectors. The neuron whose weight vector has the highest similarity is called the Best Matching Unit (BMU). The SOM algorithm is based on competitive learning, which means that there is only one BMU each time. (d) The neighborhood of the BMU is calculated. (e) The weight vectors of the neurons in the BMU's neighborhood (including the BMU itself) are adjusted towards the input data sample. (f) Stop the algorithm if the convergence condition is valid, otherwise go back to step (b). For a more insightful discussion on SOM, please refer to [43, 44].

1.3 Semi-supervised Learning

Semi-supervised learning [45, 46] is a type of learning which uses both labeled and unlabeled data. Semi-supervised learning is useful for a few reasons. First, in many real-world applications, the acquisition of labeled data is expensive and difficult while acquiring a large amount of unlabeled data is relatively easy and cheap. Second, effective use of unlabeled data during the training process actually tends to improve the performance of the trained model. In order to make the best use of unlabeled data, assumptions have to be hold in semi-supervised learning, such as smoothness assumption, cluster assumption, low-density separation assumption, and manifold assumption. Pseudo Labeling [47, 48] is a simple and efficient semi-supervised learning technique. The main idea of Pseudo Labeling is simple. Firstly, use the labeled data to train a model. Then, use the trained model to predict pseudo labels of the unlabeled data. Finally, combine the labeled data and the newly pseudo-labeled data to train the model again. There are other semi-supervised learning methods, such as Expectation Maximization (EM), co-training, transductive SVM and graph-based methods. Different methods rely on different assumptions [49]. For example, EM builds on cluster assumption, transductive SVM builds on low-density separation assumption, while graph-based methods build on the manifold assumption.

References

1. S. B. Kotsiantis, I. Zaharakis, and P. Pintelas, "Supervised machine learning: A review of classification techniques," *Emerging Artificial Intelligence Applications in Computer Engineering*, vol. 160, pp. 3–24, 2007.
2. J. Friedman, T. Hastie, and R. Tibshirani, *The Elements of Statistical Learning*. Springer Series in Statistics New York, 2001, vol. 1.
3. T. Cover and P. Hart, "Nearest neighbor pattern classification," *IEEE Trans. Information Theory*, vol. 13, no. 1, pp. 21–27, Jan. 1967.
4. L. Breiman, J. Friedman, C. J. Stone, and R. A. Olshen, *Classification and Regression Trees*. CRC Press, 1984.
5. J. Han, J. Pei, and M. Kamber, *Data Mining: Concepts and Techniques*. Elsevier, 2011.
6. J. R. Quinlan, "Induction of decision trees," *Machine Learning*, vol. 1, no. 1, pp. 81–106, 1986.
7. S. Karatsiolis and C. N. Schizas, "Region based support vector machine algorithm for medical diagnosis on Pima Indian Diabetes dataset," in *Proc. IEEE BIBE'12, Larnaca, Cyprus*, Nov. 2012, pp. 139–144.

8. W. R. Burrows, M. Benjamin, S. Beauchamp, E. R. Lord, D. McCollor, and B. Thomson, "CART decision-tree statistical analysis and prediction of summer season maximum surface ozone for the Vancouver, Montreal, and Atlantic regions of Canada," *Journal of Applied Meteorology*, vol. 34, no. 8, pp. 1848–1862, 1995.

9. A. Kumar, P. Bhatia, A. Goel, and S. Kole, "Implementation and comparison of decision tree based algorithms," *International Journal of Innovations & Advancement in Computer Science*, vol. 4, pp. 190–196, May. 2015.

10. L. Breiman, "Random forests," *Machine Learning*, vol. 45, no. 1, pp. 5–32, 2001.

11. S. Haykin, *Neural Networks: A Comprehensive Foundation*. Prentice Hall PTR, 1994.

12. S. Haykin and N. Network, "A comprehensive foundation," *Neural Networks*, vol. 2, no. 2004, p. 41, 2004.

13. K. Lee, D. Booth, and P. Alam, "A comparison of supervised and unsupervised neural networks in predicting bankruptcy of Korean firms," *Expert Systems with Applications*, vol. 29, no. 1, pp. 1–16, 2005.

14. S. Timotheou, "The random neural network: A survey," *The Computer Journal*, vol. 53, no. 3, pp. 251–267, March 2010.

15. S. Basterrech and G. Rubino, "A tutorial about random neural networks in supervised learning," *arXiv preprint arXiv:1609.04846*, 2016.

16. H. Bakircioglu and T. Koccak, "Survey of random neural network applications," *European Journal of Operational Research*, vol. 126, no. 2, pp. 319–330, 2000.

17. Y. LeCun, Y. Bengio, and G. Hinton, "Deep learning," *Nature*, vol. 521, no. 7553, p. 436, 2015.

18. J. Baker, "Artificial neural networks and deep learning," Feb. 2015. [Online]. Available: http://lancs.ac.uk/~bakerj1/pdfs/ANNs/Artificial_neural_networks-report.pdf

19. J. Schmidhuber, "Deep learning in neural networks: An overview," *Neural Networks*, vol. 61, pp. 85–117, 2015.

20. G. Pandey and A. Dukkipati, "Learning by stretching deep networks," in *International Conference on Machine Learning*, 2014, pp. 1719–1727.

21. A. Krizhevsky, I. Sutskever, and G. E. Hinton, "Imagenet classification with deep convolutional neural networks," in *Advances in Neural Information Processing Systems*, 2012, pp. 1097–1105.

22. C. Li, Y. Wu, X. Yuan, Z. Sun, W. Wang, X. Li, and L. Gong, "Detection and defense of DDoS attack-based on deep learning in OpenFlow-based SDN," *International Journal of Communication Systems*, 2018.

23. T. Mikolov, M. Karafiát, L. Burget, J. Černocký, and S. Khudanpur, "Recurrent neural network based language model," in *Eleventh Annual Conference of the International Speech Communication Association*, 2010.

24. H. Sak, A. Senior, and F. Beaufays, "Long short-term memory recurrent neural network architectures for large scale acoustic modeling," in *Fifteenth Annual Conference of the International Speech Communication Association*, 2014.

25. S. Hochreiter and J. Schmidhuber, "Long short-term memory," *Neural Computation*, vol. 9, no. 8, pp. 1735–1780, 1997.

26. X. Li and X. Wu, "Constructing long short-term memory based deep recurrent neural networks for large vocabulary speech recognition," in *Proc. IEEE ICASSP'15, Brisbane, QLD, Australia*, April 2015, pp. 4520–4524.

27. V. N. Vapnik and V. Vapnik, *Statistical Learning Theory*. Wiley New York, 1998, vol. 1.

28. B. Yekkehkhany, A. Safari, S. Homayouni, and M. Hasanlou, "A comparison study of different kernel functions for SVM-based classification of multi-temporal polarimetry SAR data," *The International Archives of Photogrammetry, Remote Sensing and Spatial Information Sciences*, vol. 40, no. 2, p. 281, 2014.

29. A. Patle and D. S. Chouhan, "SVM kernel functions for classification," in *Proc. IEEE ICATE'13, Mumbai, India*, Jan 2013, pp. 1–9.

30. I. Steinwart and A. Christmann, *Support Vector Machines*. Springer Science & Business Media, 2008.

31. M. Martínez-Ramón and C. Christodoulou, "Support vector machines for antenna array processing and electromagnetics," *Synthesis Lectures on Computational Electromagnetics*, vol. 1, no. 1, pp. 1–120, 2005.

32. H. Hu, Y. Wang, and J. Song, "Signal classification based on spectral correlation analysis and SVM in cognitive radio," in *Proc. IEEE AINA'08, Okinawa, Japan*, March. 2008, pp. 883–887.

33. G. E. Box and G. C. Tiao, *Bayesian Inference in Statistical Analysis*. John Wiley & Sons, 2011, vol. 40.

34. J. Bakker, "Intelligent traffic classification for detecting DDoS attacks using SDN/OpenFlow," *Victoria University of Wellington*, pp. 1–142, 2017.

35. N. Friedman, D. Geiger, and M. Goldszmidt, "Bayesian network classifiers," *Machine Learning*, vol. 29, no. 2–3, pp. 131–163, 1997.

36. F. V. Jensen, *An Introduction to Bayesian Networks*. UCL Press London, 1996, vol. 210.

37. D. Heckerman *et al.*, "A tutorial on learning with Bayesian networks," *Nato Asi Series D Behavioural And Social Sciences*, vol. 89, pp. 301–354, 1998.

38. T. D. Nielsen and F. V. Jensen, *Bayesian Networks and Decision Graphs*. Springer Science & Business Media, 2009.

39. L. R. Rabiner, "A tutorial on hidden markov models and selected applications in speech recognition," *Proceedings of the IEEE*, vol. 77, no. 2, pp. 257–286, Feb. 1989.

40. P. Holgado, V. A. VILLAGRA, and L. Vazquez, "Real-time multistep attack prediction based on hidden markov models," *IEEE Trans. Dependable and Secure Computing*, vol. PP, no. 99, pp. 1–1, 2017.

41. E. Alpaydin, *Introduction to Machine Learning*. MIT Press, 2014.

42. T. Kanungo, D. M. Mount, N. S. Netanyahu, C. D. Piatko, R. Silverman, and A. Y. Wu, "An efficient k-means clustering algorithm: Analysis and implementation," *IEEE Trans. Pattern Analysis and Machine Intelligence*, vol. 24, no. 7, pp. 881–892, Jul. 2002.

43. T. Kohonen, "The self-organizing map," *Neurocomputing*, vol. 21, no. 1–3, pp. 1–6, 1998.

44. M. M. Van Hulle, "Self-organizing maps," in *Handbook of Natural Computing*. Springer, 2012, pp. 585–622.

45. X. Zhu, "Semi-supervised learning literature survey," *Citeseer*, pp. 1–59, 2005.

46. X. Zhou and M. Belkin, "Semi-supervised learning," in *Academic Press Library in Signal Processing*. Elsevier, 2014, vol. 1, pp. 1239–1269.

47. D.-H. Lee, "Pseudo-label: The simple and efficient semi-supervised learning method for deep neural networks," in *Workshop on Challenges in Representation Learning, ICML*, vol. 3, 2013, p. 2.

48. H. Wu and S. Prasad, "Semi-supervised deep learning using Pseudo labels for hyperspectral image classification," *IEEE Trans. Image Processing*, vol. 27, no. 3, pp. 1259–1270, March 2018.

49. O. Chapelle, B. Scholkopf, and A. Zien, "Semi-supervised learning (chapelle, o. et al., eds.; 2006)[book reviews]," *IEEE Trans. Neural Networks*, vol. 20, no. 3, pp. 542–542, 2009.

Chapter 2
Reinforcement Learning and Deep Reinforcement Learning

Abstract In order to better understand state-of-the-art reinforcement learning agent, deep Q-network, a brief review of reinforcement learning and Q-learning are first described. Then recent advances of deep Q-network are presented, and double deep Q-network and dueling deep Q-network that go beyond deep Q-network are also given.

2.1 Reinforcement Learning

Reinforcement learning is an important branch of machine learning, where an agent learns to take actions that would yield the most reward by interacting with the environment. Different from supervised learning, reinforcement learning cannot learn from samples provided by an experienced external supervisor. Instead, it has to operate based on its own experience despite that it faces with significant uncertainty about the environment.

Reinforcement learning is defined not characterizing learning methods, but by characterizing a learning problem. Any method that is suitable for solving that problem can be considered as a reinforcement learning method [1]. A reinforcement learning problem can be described as an optimal control of a Markov Decision Process (MDP), however, state space, explicit transition probability and reward function are not necessarily required [2]. Therefore, reinforcement learning is powerful in handling tough situations that approach real-world complexity [3].

There are two outstanding features of reinforcement learning: trial-and-error search and delayed reward. Trial-and-error search means making trade-off between exploration and exploitation. The agent prefers to exploit the effective actions that have been tried in the past to produce rewards, but it also has to explore better new actions that may yield higher rewards in the future. The agent must try various actions and progressively favor those that earn the most rewards. The other feature of reinforcement learning is that the agent looks into a big picture, not just considering the immediate reward, but also the cumulative rewards in the long run, which is specified as value function.

© The Author(s), under exclusive license to Springer Nature Switzerland AG 2019   15
F. R. Yu, Y. He, *Deep Reinforcement Learning for Wireless Networks*, SpringerBriefs
in Electrical and Computer Engineering, https://doi.org/10.1007/978-3-030-10546-4_2

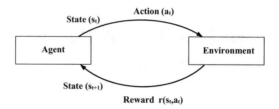

Fig. 2.1 A basic diagram of a RL system. The agent takes an action according to the current state and then receives a reward. $r(s_t, a_t)$ denotes the immediate reward that the agent receives after performing an action a_t at the state s_t

Generally, reinforcement learning can be divided into model-free and model-based reinforcement learning, which is based on whether or not the environment elements are already known. Model-free reinforcement learning has recently been successfully applied to handle the deep neural network and value functions [3–6]. It can learn policies for tough tasks using the raw state representation directly as the input to the neural networks [7]. Contrastively, model-based reinforcement learn a model of the system with the help of supervised learning and optimize a policy under this model [7–9]. Recently, elements of model-based RL have been incorporated into model-free deep reinforcement learning to accelerate the learning rate without losing the strengths of model-free learning [7]. One of the model-free reinforcement learning algorithms is Q-Learning. The most important component of Q-learning algorithms is a method for properly and efficiently estimating the Q value. The Q function can be implemented simply by a look-up table, or by a function approximator, sometimes a nonlinear approximator, such as a neural network, or even more complicated deep neural network. Q-learning, combined with deep neural network, is the so-called deep Q-learning.

2.2   Deep Q-Learning

The agent that uses a neural network to represent Q function is called Q-network, which is denoted as $Q(x, a; \theta)$. The parameter θ stands for the weights of the neural network, and the Q-network is trained by updating θ at each iteration to approximate the real Q values. While neural networks allow for great flexibility, they do so at the cost of stability when it comes to Q-Learning, which is interpreted in [3]. Deep Q-network that uses deep neural networks instead of approximating the Q-function is proposed recently, and it is proven to be more advantageous with greater performance and more robust learning [3]. To transform an ordinary Q-network into a deep Q-network, three improvements have been implemented.

- Replacing ordinary neural networks with advanced multi-layer deep convolutional networks, which utilize hierarchical layers of tiled convolutional filters to exploit the local spatial correlations, and make it possible to extract high-level features from raw input data [2, 10].

- Utilizing Experience Replay, which stores its interaction experience tuple $e(t) = (x(t), a(t), r(t), x(t + 1))$ at time t into a replay memory $D(t) = \{e(1), \ldots, e(t)\}$, and then randomly samples batches from the experience pool to train the deep convolutional network's parameters rather than directly using the consecutive samples as in Q-learning. This allows the network to learn from more various past experiences, and restrains the network from only focusing on what it is immediately doing.

- Adopting a second network, to generate the target Q values that are used to calculate the loss for each action during the training procedure. One network for both estimations of Q values and target Q values would result in falling into feedback loops between the target and estimated values. Thus, in order to stabilize the training, the target network's weights are fixed and periodically updated.

The deep Q-function is trained towards the target value by minimizing the loss function $L(\theta)$ at each iteration, and the loss function can be written as

$$L_i(\theta_i) = E[(y_i - Q(x, a; \theta_i)^2)], \tag{2.1}$$

where $y_i = r + \varepsilon \max_{a'} Q(x', a'; \theta_i^-)$. Here, the weights θ_i^- are updated as $\theta_i^- = \theta_{i-G}$, i.e., the weights update every G time steps in the deep Q-networks instead of $\theta_i^- = \theta_{i-1}$.

2.3   Beyond Deep Q-Learning

Since deep Q-learning is first proposed, great efforts have been made for even greater performance and higher stability. Here, we briefly introduce two recent improvements: Double DQN [11] and Dueling DQN [4].

2.3.1   Double DQN

In regular DQN, both choosing an action and evaluating the chosen action use the max over the Q values, which would lead to over-optimistic Q value estimation. To relieve the over-estimation problem, the target value in double DQN is designed and updated as

$$y_i^{double} = r + \varepsilon Q(x', \arg \max_{a'} Q(x', a'; \theta_i); \theta_i^-), \tag{2.2}$$

where the action choice is decoupled from the target Q-value generation. This simple trick makes the over-estimation significantly reduced, and the training procedure runs faster and more reliably.

2.3.2 Dueling DQN

The intuition behind Dueling DQN is that it is not always necessary to estimate the value of taking each available action. For some states, the choice of action makes no influence on what happens. Thus, in dueling DQN, the state-action value $Q(x, a)$ is decomposed into two components as follows,

$$Q(x, a) = V(x) + A(a). \qquad (2.3)$$

Here, $V(x)$ is the value function, which simply represents how good it is to be in a given state x. $A(a)$ is the advantage function, which measures the relative importance of a certain action compared with other actions. After $V(x)$ and $A(a)$ are separately computed, their values are combined back into a single Q-function at the final layer. This improvement would lead to better policy evaluation.

Combining both the above two techniques can achieve better performance and faster training speed. In the following section, the deep Q-network with the two improvements is utilized to find the optimal policy for a practical network application.

References

1. R. S. Sutton, "Introduction: The challenge of reinforcement learning," in *Reinforcement Learning*. Springer, 1992, pp. 1–3.
2. H. Y. Ong, K. Chavez, and A. Hong, "Distributed deep Q-learning," *arXiv preprint arXiv:1508.04186*, 2015.
3. V. Mnih, K. Kavukcuoglu, D. Silver, A. A. Rusu, J. Veness, M. G. Bellemare, A. Graves, M. Riedmiller, A. K. Fidjeland, G. Ostrovski *et al.*, "Human-level control through deep reinforcement learning," *Nature*, vol. 518, no. 7540, pp. 529–533, 2015.
4. Z. Wang, N. de Freitas, and M. Lanctot, "Dueling network architectures for deep reinforcement learning," *arXiv preprint arXiv:1511.06581*, 2015.
5. M. Hausknecht and P. Stone, "Deep reinforcement learning in parameterized action space," *arXiv preprint arXiv:1511.04143*, 2015.
6. T. P. Lillicrap, J. J. Hunt, A. Pritzel, N. Heess, T. Erez, Y. Tassa, D. Silver, and D. Wierstra, "Continuous control with deep reinforcement learning," *arXiv preprint arXiv:1509.02971*, 2015.
7. S. Gu, T. Lillicrap, I. Sutskever, and S. Levine, "Continuous deep Q-learning with model-based acceleration," *arXiv preprint arXiv:1603.00748*, 2016.
8. S. Levine, C. Finn, T. Darrell, and P. Abbeel, "End-to-end training of deep visuomotor policies," *Journal of Machine Learning Research*, vol. 17, no. 39, pp. 1–40, 2016.

9. M. Deisenroth and C. E. Rasmussen, "Pilco: A model-based and data-efficient approach to policy search," in *Proceedings of the 28th International Conference on machine learning (ICML-11)*, 2011, pp. 465–472.
10. V. Mnih, K. Kavukcuoglu, D. Silver, A. Graves, I. Antonoglou, D. Wierstra, and M. Riedmiller, "Playing atari with deep reinforcement learning," *arXiv preprint arXiv:1312.5602*, 2013.
11. H. Van Hasselt, A. Guez, and D. Silver, "Deep reinforcement learning with double q-learning," *CoRR, abs/1509.06461*, 2015.

Chapter 3
Deep Reinforcement Learning for Interference Alignment Wireless Networks

Abstract Both caching and interference alignment (IA) are promising techniques for next generation wireless networks. Nevertheless, most existing works on cache-enabled IA wireless networks assume that the channel is invariant, which is unrealistic considering the time-varying nature of practical wireless environments. In this chapter, we consider realistic time-varying channels. Specifically, the channel is formulated as a finite-state Markov channel (FSMC). The complexity of the system is very high when we consider realistic FSMC models. Therefore, in this chapter, we propose a novel deep reinforcement learning approach, which is an advanced reinforcement learning algorithm that uses deep Q network to approximate the Q value-action function. We use Google TensorFlow to implement deep reinforcement learning in this chapter to obtain the optimal IA user selection policy in cache-enabled opportunistic IA wireless networks. Simulation results are presented to show that the performance of cache-enabled opportunistic IA networks in terms of the network's sum rate and energy efficiency can be significantly improved by using the proposed approach.

3.1 Introduction

Recently, *wireless proactive caching* has attracted great attentions from both academia and industry [1, 2]. By effectively reducing the duplicate content transmissions in networks, caching has been recognized as one of the promising techniques for future wireless networks to improve spectral efficiency, shorten latency, and reduce energy consumption [3–5]. Based on the global traffic features, a few popular contents are requested by many users during a short time span, which accounts for most of the traffic load. Therefore, proactively caching the popular contents can remove the heavy burden of the backhaul links.

Another new technology called *interference alignment* (IA) has been studied extensively as a revolutionary technique to tackle the interference issue in wireless networks [6]. IA exploits the cooperation of transmitters to design the precoding matrices, and thus eliminates the interferences. IA can benefit mobile cellular networks [7]. Due to the large number of users in cellular networks, multiuser

© The Author(s), under exclusive license to Springer Nature Switzerland AG 2019   21
F. R. Yu, Y. He, *Deep Reinforcement Learning for Wireless Networks*, SpringerBriefs
in Electrical and Computer Engineering, https://doi.org/10.1007/978-3-030-10546-4_3

diversity has been studied in conjunction with IA, called opportunistic IA, which further improves the network performance [8–11].

Jointly considering these two important technologies, caching and IA, can be beneficial in IA-based wireless networks [12, 13]. The implementation of IA requires the channel state information (CSI) exchange among transmitters, which usually relies on the backhaul link. The limited capacity of backhaul link has significant impacts on the performance of IA [14]. Caching can relieve the traffic loads of backhaul links, thus the saved capacity can be used for CSI exchange in IA. In [12], the authors investigate the benefits of caching and IA in the context of multiple-input and multiple-output (MIMO) interference channels, and maximize the average transmission rate by optimizing the number of the active transceiver pairs. In [13], it is shown that by properly placing the content in the transmitters' caches, the IA gain can be increased.

Although some excellent works have been done on caching and IA, most of these previous works assume that the channel is block-fading channel or invariant channel, where the estimated CSI of the current time instant is simply taken as the predicted CSI for the next time instant. Considering the time-varying nature of wireless environments, this kind of memoryless channel assumption is not realistic[15], especially in vehicular environments due to high mobility [16]. In addition, it is difficult to obtain the perfect CSI due to channel estimation errors, communication latency and backhaul link constraints [17, 18].

In this chapter, we consider realistic time-varying channels, and propose a novel *deep reinforcement learning* approach in cache-enabled opportunistic IA wireless networks. Deep reinforcement learning is a kind of machine learning, which is a powerful tool to process wireless data[19]. The distinct features of this chapter are as follows.

- Cache-enabled opportunistic IA is studied under the condition of time-varying channel coefficients. The channel is formulated as a finite-state Markov channel (FSMC), which has been widely accepted in the literature to characterize the correlation structure of the fading process [20–22]. Considering FSMC models may enable significant performance improvements over the schemes with memoryless channel models.
- The complexity of the system is very high when we consider realistic FSMC models. Therefore, we propose a novel deep reinforcement learning approach in this chapter. Deep reinforcement learning is an advanced reinforcement learning algorithm that uses deep Q network to approximate the Q value-action function [23], and it has been used in wireless networks to improve the performance[24, 25]. Deep reinforcement learning is used in this chapter to obtain the optimal IA user selection policy in cache-enabled opportunistic IA wireless networks.
- We use Google TensorFlow to implement deep reinforcement learning. The visualization of the deep Q network model is presented. Simulation results with different system parameters are presented to show the effectiveness of the proposed scheme. It is illustrated that the performance of cache-enabled opportunistic IA

networks in terms of the network's sum rate and energy efficiency can be significantly improved by using the proposed deep reinforcement learning approach.

The rest of this chapter is organized as follows. In Sect. 3.2, the system model is presented. The deep reinforcement learning algorithm is presented in Sect. 3.3. In Sect. 3.3, the cache-enabled opportunistic IA network is formulated as a deep reinforcement learning process. Simulation results are discussed in Sect. 3.4. Finally, conclusions are presented in Sect. 3.5.

3.2 System Model

In this section, we describe the model of IA, followed by the time-varying channel. Then, cache-equipped transmitters are described.

3.2.1 Interference Alignment

IA is a revolutionary interference management technique, which theoretically enables the network's sum rate grow linearly with the cooperative transmitter and receiver pairs. That is to say, each user can obtain the capacity $\frac{1}{2}\log(\text{SNR}) + o(\log(\text{SNR}))$, which has nothing to do with the interferences [26]. Actually, SNR plays a crucial role in determining the IA results. Cadambe and Jafar pointed out that IA performs better at very high SNR, and suffers from low quality at moderate SNR levels [26]. Meanwhile higher and higher SNR is required to approach IA network's theoretical maximum sum rate as the number of IA users increases [6]. Thus, there exist competitions among users for accessing to IA network.

Consider a K-user MIMO interference channel. $N_t^{[k]}$ and $N_r^{[k]}$ antennas are equipped at the kth transmitter and receiver, respectively. The degree of freedom (DoF) of the kth user is denoted as $d^{[k]}$. The received signal at the kth receiver can be written as

$$\mathbf{y}^{[k]}(t) = \mathbf{U}^{[k]\dagger}(t)\mathbf{H}^{[kk]}(t)\mathbf{V}^{[k]}(t)\mathbf{x}^{[k]}(t)$$

$$+ \sum_{j=1, j \neq k}^{K} \mathbf{U}^{[k]\dagger}(t)\mathbf{H}^{[kj]}(t)\mathbf{V}^{[j]}(t)\mathbf{x}^{[j]}(t) + \mathbf{U}^{[k]\dagger}(t)\mathbf{z}^{[k]}(t), \tag{3.1}$$

where the first term at the right side represents the expected signal, and the other two terms mean the inter-user interference and noise, respectively. $\mathbf{H}^{[kj]}(t)$ is the $N_r^{[k]} \times N_t^{[j]}$ matrix of channel coefficients from the jth transmitter to the kth receiver over the time slot t. Each element of $\mathbf{H}^{[kj]}(t)$ is independent and identically distributed (i.i.d) complex Gaussian random variable, with zero mean and unit variance. $\mathbf{V}^{[k]}(t)$ and $\mathbf{U}^{[k]}(t)$ are the unitary $N_t^{[k]} \times d^{[k]}$ precoding matrix and $N_r^{[k]} \times d^{[k]}$ interference suppression matrix of the kth user, respectively. $\mathbf{x}^{[k]}(t)$ and $\mathbf{z}^{[k]}(t)$ are the transmitted signal vector of $d^{[k]}$ dimensions and the $N^{[k]} \times 1$

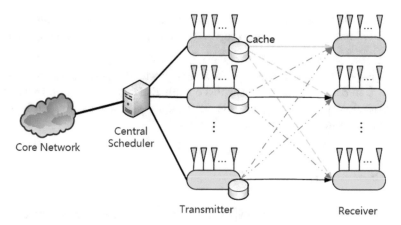

Fig. 3.1 System model of a cache-enabled opportunistic IA wireless network

additive white Gaussian noise (AWGN) vector whose elements have zero mean and σ^2 variance at the kth receiver, respectively.

The interference can be perfectly eliminated only when the following conditions can be satisfied

$$\mathbf{U}^{[k]\dagger}(t)\mathbf{H}^{[kj]}(t)\mathbf{V}^{[j]}(t) = 0, \ \forall j \neq k, \tag{3.2}$$

$$\text{rank}\left(\mathbf{U}^{[k]\dagger}(t)\mathbf{H}^{[kk]}(t)\mathbf{V}^{[k]}(t)\right) = d^{[k]}. \tag{3.3}$$

Under this assumption, the received signal at the kth receiver can be rewritten as

$$\mathbf{y}^{[k]}(t) = \mathbf{U}^{[k]\dagger}(t)\mathbf{H}^{[kk]}(t)\mathbf{V}^{[k]}(t)\mathbf{x}^{[k]}(t) + \mathbf{U}^{[k]\dagger}(t)\mathbf{z}^{[k]}(t). \tag{3.4}$$

To meet Condition (3), the global CSI is required at each transmitter. Each transmitter can estimate its local CSI (i.e., the direct link), but the CSI of other links can only be obtained by CSI share with other transmitters via the backhaul link[12]. Thus, in IA network, the backhaul link is more than a pipeline for connecting with Internet. The limited capacity should be made optimum use of. The recent advances focus on the benefits of edge caching, which is capable to decrease the data transfer and leave more capacity for CSI share. The detail is described in the following subsection. In this chapter, we assume the total backhaul link capacity of all the users is C_{total}, and the CSI estimation is perfect with no errors and no time delay.

3.2.2 Cache-Equipped Transmitters

In the era of explosive information, the vast amount of content makes it impossible for all of them gain popularity. As a matter of fact, only a small fraction becomes extensively popular. That means certain content may be requested over and over during a short time span, which gives rise to the network congestion and transmission delay. We assume that each transmitter is equipped with a cache unit that has certain amount of storage space. The stored content may follow a certain popularity distribution.

For consistency, the cache of each transmitter stores the same content, usually the web content, and thus alleviating the backhaul burden and shorten delay time. In [27], the authors survey on the existing methods for predicting the popularity of different types of web content. Specifically, they show that different types of content follow different popularity distributions. For example, the popularity growth of online videos complies with power-law or exponential distributions, that of the online news can be represented by power-law or log-normal distributions, etc. Based on the content popularity distribution and cache size, cache hit probability P_{hit} and cache miss probability P_{miss} can be derived[12]. In this chapter, the specific popularity distribution is not the focus, and we just concentrate on two states, whether the requested content is within the cache or not. We describe the two states as $\Lambda = \{0, 1\}$, where 0 means the requested content is not within the cache, and 1 indicates it is within the cache.

In this chapter, we consider an MIMO interference network with limited backhaul capacity and caches equipped at the transmitter side, as illustrated in Fig. 3.1. There is a central scheduler who is responsible for collecting the channel state and cache status from each user, scheduling the users and allocating the limited resources. All the users are connected to the central scheduler via a backhaul link for CSI exchange and Internet connection, and the total capacity is limited.

When we consider realistic FSMC models, the complexity of the system is very high, which makes it difficult to be solved using traditional approaches. In this chapter, we used a novel reinforcement learning approach to solve the optimization problem in cache-enabled opportunistic IA wireless networks.

3.3 Problem Formulation

In this section, we first formulate the realistic time-varying channels as finite-state Markov channels (FSMC), and then we demonstrate how to formulate the cache-enabled IA network optimization problem as a deep Q-learning process, which can determine the optimal policy for IA user grouping.

3.3.1 Time-Varying IA-Based Channels

We model the realistic time-varying channels as FSMCs, which is an effective method to characterize the fading nature of wireless channels[20]. Specifically, the first-order FSMC is used in this chapter.

Here we consider a multi-user interference network with one DoF for each user, and the received signal to interference and noise ratio (SINR) is an important parameter that can be used to reflect the quality of a wireless channel. In an IA-based wireless network, if the interference is completely eliminated, the received SINR of the kth user at time instant t can be derived as follows,

$$\text{SINR}^{[k]}(t) = \frac{|h^{[kk]}(t)|^2 P^{[k]}}{\sigma^2}, \tag{3.5}$$

where $h^{[kk]}(t) = \mathbf{u}^{[k]\dagger}(t)\mathbf{H}^{[kk]}(t)\mathbf{v}^{[k]}(t)$. On the other hand, if the interference is not perfectly eliminated, the received SINR is

$$\text{SINR}^{[k]}(t) = \frac{|h^{[kk]}(t)|^2 P^{[k]}}{\sum_{j \neq k} |h^{[kj]}|^2 P^{[j]} + \sigma^2}, \tag{3.6}$$

where $h^{[kj]}(t) = \mathbf{u}^{[k]\dagger}(t)\mathbf{H}^{[kj]}(t)\mathbf{v}^{[j]}(t)$.

As we consider opportunistic IA networks with user scheduling, directly modeling the received SINR of a certain user as a Markov random variable is not appropriate. Due to the relationship of the received SINR and the channel coefficient, we model the channel coefficient after using IA, $|h^{[kj]}|^2$, as a Markov random variable. Here, $|h^{[kj]}(t)|^2$ is newly modelled because of its unknown distribution. We partition and quantize the range of $|h^{[kj]}|^2$ into H levels. Each level corresponds to a state of the Markov channel. This channel information is included in the system state, which will be discussed in detail in the following subsection. We consider T time slots over a period of wireless communications. Let us denote $t \in \{0, 1, 2, \ldots, T - 1\}$ as the time instant, and the channel coefficient varies from one state to another state when one time slot elapses.

3.3.2   Formulation of the Network's Optimization Problem

In our system, there are L candidates that want to join in the IA network to communicate wirelessly. We assume that the IA network size is always smaller than the number of candidates, which is in accordance with the fact that a large number of users expect wireless communications anytime and anywhere. As aforementioned, the value of SNR affects the performance of interference alignment, and the candidates who occupy the better channels are more advantageous for accessing to the IA network. Therefore, we make an action at each time slot to decide which candidates are the optimal users for constructing an IA network based on their current states.

Here, a central scheduler is responsible for acquiring each candidate's CSI and cache status, then it assembles the collected information into a system state. Next, the controller sends the system state to the agent, i.e., the deep Q network, and then the deep Q network feeds back the optimal action $\arg\max_{\pi} Q^*(x, a)$ for the current time instant. After obtaining the action, the central scheduler will send a bit to inform the users to be active or not, and the corresponding precoding vector will be sent to each active transmitter. The system will transfer to a new state after an action is performed, and the rewards can be obtained according to the reward function.

Algorithm 1 Deep reinforcement learning algorithm in cache-enabled IA networks

---

Initialization.

Initialize replay memory.
Initialize the Q network parameters with θ.
Initialize the target Q network parameters with $\theta^- = \theta$.

For episode $k = 1, \ldots, K$ **do**:

Initialize the beginning state x.

 For $t = 1, 2, 3 \ldots, T$ **do**:

   Choose a random probability p.
   Choose $a(t)$ as,
     if $p \geq \varepsilon$,
       $a^*(t) = \arg\max_a Q(x, a, \theta)$,
     otherwise,
       randomly choose a solution $a(t) \neq a^*(t)$,
   Execute $a(t)$ in the system. Observe the reward $r(t)$, and next state $x(t + 1)$. Store the
 experience $(x(t), a(t), r(t), x(t + 1))$ in replay memory.
   Get mini-batch of samples $(x(t), a(t), \zeta(t), x(t + 1))$ from the replay memory.
   Perform a gradient descent step on $(y - Q(x, a, \theta)^2)$ with respect to θ.
   Return the value of parameters θ in the deep Q network.

 End for
 End for

Inside the deep Q network, the replay memory stores the agent's experience of each time slot. The Q network parameter θ is updated at every time instant with samples from the replay memory. The target Q network parameter θ^- is copied from the Q network every N time instants. The ε-greedy policy is utilized to balance the exploration and exploitation, i.e., to balance the reward maximization based on the knowledge already known with trying new actions to obtain knowledge unknown. The training algorithm of the deep Q network is described in Algorithm 1.

In order to obtain the optimal policy, it is necessary to identify the actions, states and reward functions in our deep Q learning model, which will be described in the next following subsections.

System State

The current system state $x(t)$ is jointly determined by the states of L candidates. The system state at time slot t is defined as,

$$x(t) = \{|h^{[11]}(t)|^2, |h^{[12]}(t)|^2, \ldots, |h^{[kj]}(t)|^2, \ldots, |h^{[LL]}(t)|^2,$$
$$c_1(t), c_2(t), \ldots, c_l(t), \ldots, c_L(t)\}. \tag{3.7}$$

Here, there are two components in the state: the channel coefficient after being processed by IA technique $|h^{[kj]}(t)|^2$ and the cache state $c_i(t)$. The cache state $c_l(t) \in \Lambda = \{0, 1\}$, the index l means the lth candidate, and $l = 1, 2, \ldots, L$.

The number of possible system states can be very large. Due to the curse of dimensionality, it is difficult for traditional approaches to handle our problem. Fortunately, deep Q network is capable of successfully learning directly from high-dimensional inputs[23], thus it is proper to be used in our system.

System Action

In the system, the central scheduler has to decide which candidates to be set active, and the corresponding resources will be allocated to the active users.

The current composite action $a(t)$ is denoted by

$$a(t) = \{a_1(t), a_2(t), \ldots, a_L(t)\}, \tag{3.8}$$

where $a_l(t)$ represents the control of the lth candidate, i.e., $a_l(t) = 0$ means the candidate l is passive (not selected) at time slot t, and $a_l(t) = 1$ means it is active (selected). Due to the constraint of IA, the condition $\sum_{l=1}^{L} a_l(t) \geq 3$ must be satisfied.

Reward Function

The system reward represents the optimization objective, and we take the objective to maximize the IA network's throughput, and the reward function of the lth candidate is defined as,

$$r_l(t) =$$

$$
\begin{cases}
a_l(t) \log_2 \left(1 + \dfrac{|h^{[ll]}(t)|^2 P^{[l]} x_l}{\sum\limits_{j=1, j \neq l}^{L} a_j(t) |h^{[lj]}(t)|^2 P^{[j]} x_j + \sigma^2} \right), & \\
\qquad\qquad \text{if } c_l(t) = 1, & \\[2em]
a_l(t) \min \left\{ \left[\dfrac{1}{\sum\limits_{i=1}^{L} a_i(t)} (C_{total} - C_c \sum\limits_{i=1}^{L} a_i(t)) \right], \right. & \\
\qquad \left. \log_2 \left(1 + \dfrac{|h^{[ll]}(t)|^2 P^{[l]} x_l}{\sum\limits_{j=1, j \neq l}^{L} a_j(t) |h^{[lj]}(t)|^2 P^{[j]} x_j + \sigma^2} \right) \right\} & \\
\qquad\qquad \text{if } c_l(t) = 0, &
\end{cases}
\tag{3.9}
$$

where $\sum\limits_{i=1}^{L} a_i(t) \geq 3$ is a condition for IA network, C_{total} is the total capacity of the backhaul link, and C_c is the reserved capacity allocated to each active user to share CSI with other active users. For the lth candidate, if the requested content is not in the local cache, it can only acquire the content through the backhaul link, and equal capacity (the total capacity minus the total capacity for CSI exchange) is allocated among the active users. If the requested content is within the cache, the lth candidate can get the maximum rate that an IA user can achieve. We assume the realistic IA situations that the interference cannot be perfectly eliminated, i.e., the interference leakage from other non-direct channels remains. Thus, the reward of the lth candidate is determined by the integrated system's state, including the states of both the direct and non-direct links.

The immediate system reward is the sum of all the candidates' immediate rewards, i.e., $r(t) = \sum\limits_{n=1}^{L} r_n(t)$. The central scheduler gets $r(t)$ in state $x(t)$ when action $a(t)$ is performed in time slot t. The goal of using deep Q network into our system model is to find a selection policy that maximizes the discounted cumulative rewards during the communication period T, and the cumulative reward can be expressed as

$$R = \max_{\pi} E\left[\sum_{t=0}^{T-1} \epsilon^t r(t)\right], \tag{3.10}$$

where ϵ^t approaches to zero when t is large enough. In practice, a threshold for terminating the process can be set.

The Effects of Imperfect CSI

In (3.7) and (3.9), we assume perfect CSI. However, in practical networks, the implementation of IA requires the CSI exchange among transmitters on the backhaul link, which will cause the delay of CSI. The limited capacity of backhaul link can have dramatic effects on the performance of IA due to the delayed CSI. Suppose that $\mathbf{H}(t)$ is the matrix consisting of the accurate channel coefficients at time instant t. The delayed channel matrix is denoted as follows.

$\mathbf{H}(t - \tau)$ is the delayed channel matrix when CSI is delayed by τ duration. The relation between the delayed CSI and the current CSI can be modeled as

$$\mathbf{H}(t) = \rho\mathbf{H}(t - \tau) + \sqrt{1 - \rho^2}\delta = \rho\widehat{\mathbf{H}}(t) + \sqrt{1 - \rho^2}\delta, \tag{3.11}$$

where $\widehat{\mathbf{H}}(t)$ is the matrix of the delayed channel coefficients at the time instant t. δ has the same distribution with $\mathbf{H}(t)$ and $\widehat{\mathbf{H}}(t)$. ρ is the normalized autocorrelation function of a fading channel with motion at a constant velocity, and $0 \leq \rho \leq 1$. It can be seen that $\rho = 1$ corresponds to perfect CSI, whereas $\rho = 0$ represents no CSI. ρ is defined as follows.

$$\rho = \frac{\mathbb{E}\left[(\mathbf{H})_{ij}\left(\widehat{\mathbf{H}}\right)^{*}_{ij}\right]}{\sqrt{\mathbb{E}\left[|(\mathbf{H})_{ij}|^{2}\right]\mathbb{E}\left[\left|(\widehat{\mathbf{H}})_{ij}\right|^{2}\right]}}. \tag{3.12}$$

The value of ρ depends on the time scale of channel variation, which can be defined by coherence time [28], and length of the delay τ. When the channel is under Rayleigh fading, ρ can be derived as follows [29]

$$\rho = J_0(2\pi f_d \tau), \tag{3.13}$$

where $J_0(.)$ is the zeroth order Bessel function of the first kind, and f_d is the Doppler frequency, which reflects the velocity of the transceivers.

Derivation of $\mathbf{u}$ and $\mathbf{v}$

Here we leverage the conventional iterative interference alignment[26] to derive the precoding vector $\mathbf{v}$ and decoding vector $\mathbf{u}$ for each user, which utilizes the reciprocity of wireless channels and aims at minimizing the total interference leakage at receivers.

In the forward direction of iterations, the total interference leakage at the lth receiver caused by other users can be written as

$$I^{[l]} = \mathrm{Tr}\left[\mathbf{u}^{[l]\dagger}\mathbf{Q}^{[l]}\mathbf{u}^{[l]}\right], \tag{3.14}$$

where Tr[$\mathbf{A}$] denotes the trace operator of matrix $\mathbf{A}$, and

$$\mathbf{Q}^{[l]} \triangleq \sum_{i=1,i\neq l}^{L} P^{[i]}\mathbf{H}^{[li]}\mathbf{v}^{[i]}\mathbf{v}^{[i]\dagger}\mathbf{H}^{[li]\dagger}. \tag{3.15}$$

In the reverse direction of iterations, the total interference leakage at the ith receiver (i.e., the ith transmitter in the original direction) can be denoted as

$$\overleftarrow{I}^{[i]} = \mathrm{Tr}\left[\overleftarrow{\mathbf{u}}^{[i]\dagger}\overleftarrow{\mathbf{Q}}^{[i]}\overleftarrow{\mathbf{u}}^{[i]}\right], \tag{3.16}$$

where

$$\overleftarrow{\mathbf{Q}}^{[i]} = \sum_{l=1,l\neq i}^{L} P^{[l]}\overleftarrow{\mathbf{H}}^{[il]}\overleftarrow{\mathbf{v}}^{[l]}\overleftarrow{\mathbf{v}}^{[l]\dagger}\overleftarrow{\mathbf{H}}^{[il]\dagger}. \tag{3.17}$$

In (3.16) and (3.17), $\overleftarrow{\mathbf{v}}^{[l]} = \mathbf{u}^{[l]}$, $\overleftarrow{\mathbf{u}}^{[l]} = \mathbf{v}^{[l]}$, and $\overleftarrow{\mathbf{H}}^{[li]} = \mathbf{H}^{[il]\dagger}$.

In the original channel, the decoding vector $\mathbf{u}^{[l]}$ of the lth user, which can minimize the total interference leakage, can be updated as

$$\mathbf{u}^{[l]} = \nu_i\left(\mathbf{Q}^{[l]}\right). \tag{3.18}$$

where $\nu_i(\mathbf{X})$ means the eigenvector corresponding to the ith smallest eigenvalue of matrix $\mathbf{X}$.

In the reverse channel, the precoding vector of the lth user is set as $\overleftarrow{\mathbf{v}}^{[l]} = \mathbf{u}^{[l]}$. The reverse decoding vector $\overleftarrow{\mathbf{u}}^{[l]}$ can be obtained as

$$\overleftarrow{\mathbf{u}}^{[l]} = \nu_i\left(\overleftarrow{\mathbf{Q}}^{[l]}\right). \tag{3.19}$$

Then the reverse decoding vector $\overleftarrow{\mathbf{u}}^{[l]}$ is regarded as the precoding vector in the original channel, i.e., $\mathbf{v}^{[l]} = \overleftarrow{\mathbf{u}}^{[l]}$. This process iterates in such a way until convergence, and the process is summarized in Algorithm 2.

3.4 Simulation Results and Discussions

Computer simulations are carried out to demonstrate the performance of the proposed deep reinforcement learning approach to the optimization of cache-enabled opportunistic IA wireless networks. We use TensorFlow [30] in our simulations

---

Algorithm 2 The conventional iterative interference alignment

---

Initialize the $N_t^{[l]} \times 1$ precoding vector $\mathbf{v}^{[l]}$ arbitrarily.

Execute the iteration

Calculate the interference covariance matrix $\mathbf{Q}^{[l]}$ for the lth receiver based on (3.15), $l = 1, 2, \ldots, L$.

Compute the decoding vector $\mathbf{u}^{[l]}$ for the lth receiver based on (3.18), $l = 1, 2, \ldots, L$.

Reverse the communication direction. Set the reversed precoding vector as the original decoding vector, i.e., $\overleftarrow{\mathbf{v}}^{[l]} = \mathbf{u}^{[l]}, l = 1, 2, \ldots, L$.

For the reversed communication direction, calculate matrix $\overleftarrow{\mathbf{Q}}^{[k]}$ for the new lth receiver based on (3.17), $l = 1, 2, \ldots, L$.

Compute the reversed decoding vector $\overleftarrow{\mathbf{u}}^{[l]}$ for the new lth receiver based on (3.19), $l = 1, 2, \ldots, L$.

Reverse the communication direction. Set the original precoding vector as the reversed decoding vector, i.e., $\mathbf{v}^{[l]} = \overleftarrow{\mathbf{u}}^{[l]}, l = 1, 2, \ldots, L$.

Continue till Convergence

Obtain the Interference alignment solutions as $\mathbf{v}^{[l]}$ and $\mathbf{u}^{[l]}, l = 1, 2, \ldots, L$.

---

to implement deep reinforcement learning. In this section, we first introduce TensorFlow, followed by simulation settings. Then, simulation results are discussed.

3.4.1 TensorFlow

TensorFlow is an open source software library for expressing machine learning algorithms, and an implementation for executing such algorithms. TensorFlow has attracted great interests from both academia and industry for a variety of applications, such as speech recognition, Gmail, Google Photos, and search [30]. Originally developed by the Google Brain team for Google's research and production purposes, TensorFlow was later released under the Apache 2.0 open source license in 2015. TensorFlow is Google Brain's second generation machine learning system to replace its predecessor DistBelief for the implementation and deployment of large-scale machine learning models.

TensorFlow provides a Python API, as well as a less documented C++ API. The reference implementation of TensorFlow runs on single devices. Nevertheless, TensorFlow can run on multiple CPUs and GPUs for fast execution. A wide variety of hardware platforms can be used for TensorFlow, which takes computations and maps them onto different hardware platforms, ranging from mobile device platforms (e.g., Android and iOS) to modest-sized systems using single machines containing one or many GPU cards to large-scale systems running hundreds and thousands of GPUs.

TensorFlow is used to transform an ordinary Q-Network into a deep Q-Network (DQN) by making the following improvements:

(1) Extending a simple neural network to a multi-layer convolutional network. We utilize the tf.contrib.layers.convolution2d function to easily create a convolutional layer as follows: $convolution\_layer = tf.contrib.layers.convolution2d$ $(in, num\_out, k\_size, stride, padding)$, where num_out is the number of filters applied to the previous layer, k_size refers to how large a window used to slide over the previous layer, $stride$ is the number of points skipped as we slide the window across the layer, and $padding$ indicates if the window just slides over the bottom layer or adds padding around it to ensure that the convolutional layer has the same dimensions as the previous layer. Please refer to the Tensorflow documentation [31] for more information.

(2) Implementing Experience Replay, which will allow our network to train itself using the stored memories from its experience. By keeping the experiences, the network can learn from a more varied sets of past experiences. We use a tuple of $< state, action, reward, nextstate >$ to store these experiences. In our DQN, a class is used to handle storing and retrieving memories.

(3) Utilizing a second "target" network, which is used to compute the target Q-values during the training procedure. The reason why two networks are used is as follows. The Q-networks values shift at every step of training, and the value

estimations can easily spiral out of control, if we adjust our network values by using a constantly shifting set of values. Consequently, there will be feedback loops between the target and estimated Q-values, and the network can become destabilized. To address this issue, we fix the target networks weights, and they are periodically updated to the primary Q-networks values. By doing this, we can have a training process in a more stable manner.

3.4.2 Simulation Settings

In our simulations, we used a GPU-based server, which has 4 Nvidia GPUs with version GTX TITAN. The CPU is Intel Xeno E5-2683 v3 with 128G memory. The software environment we utilized is TensorFlow 0.12.1 with Python 2.7 on Ubuntu 14.04 LTS.

For performance comparison, our proposed OIA scheme is compared with four other schemes as

1. The same proposed scheme without (w.o.) caching.
2. An existing selection scheme without (w.o.) caching [32], in which invariant channels are assumed, i.e., the estimated channel coefficient of the current time instant is simply taken as the predicted channel coefficient of the next time instant.
3. An existing selection scheme with (w.) caching [12], in which invariant channels are assumed, and it schedules the transceiver pairs to maximize the network throughput.
4. An existing scheme with (w.) caching [33], in which invariant channels are assumed, and power allocation strategy is performed aiming at maximizing the network throughput, however, no user selection is included.

In the simulations, we consider a cache-enabled opportunistic IA network, in which $L = 5$ candidates want to access to. Due to the feasibility of IA[34], i.e., $N_t + N_r \geq d(L + 1)$. Here, we set DoF d to be 1, and assume that each candidate is equipped with three antennas at both the transmitter node and the receiver node, i.e. $N_t = N_r = 3$. Other setup parameters are: the bandwidth $B = 10\,\text{MHz}$ per transmitter, the total backhaul capacity $C_{total} = 60\,\text{Mb/s}$, and the reserved capacity for sharing CSI $C_c = 2\,\text{Mb/s}$ per active user. The noise power σ^2 is set to be $0.1\,\text{mW}$ throughout the simulation. The normalized autocorrelation value ρ is set to be 0.99 in this chapter.

Based on the definition of system states, we quantize and uniformly partition the channel coefficients $|h^{[kj]}|^2$ into 10 levels with 9 boundary values as 10^{-6}, 0.3, 0.6, 0.9, 1.2, 1.5, 1.8, 2.1, 2.4. Uniformly setting the boundary values of FSMC is widely used in the literature (e.g., [35]) for simplicity. Nevertheless, the boundary values for a particular environment under certain criterion should be optimized for better performance [35]. We assume that the channel state transition probability is identical for all the candidates. In one simulation scenario, the transition probability

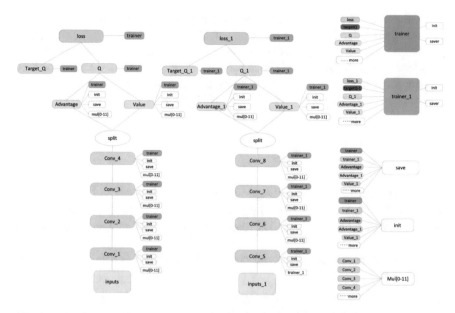

Fig. 3.2 Visualization of the proposed deep reinforcement learning algorithm using TensorBoard

of remaining in the same state is set to be 0.489, and the probability of transition to
the adjacent state to be twice that of transition to a nonadjacent state. The channel
state transition matrix is shown on the top of the next page. We change the channel
state transition probability in other simulation scenarios.

$$
P_{channel} = \begin{pmatrix}
0.489 & 0.256 & 0.128 & 0.064 & 0.032 & 0.016 & 0.008 & 0.004 & 0.002 & 0.001 \\
0.001 & 0.489 & 0.256 & 0.128 & 0.064 & 0.032 & 0.016 & 0.008 & 0.004 & 0.002 \\
0.002 & 0.001 & 0.489 & 0.256 & 0.128 & 0.064 & 0.032 & 0.016 & 0.008 & 0.004 \\
0.004 & 0.002 & 0.001 & 0.489 & 0.256 & 0.128 & 0.064 & 0.032 & 0.016 & 0.008 \\
0.008 & 0.004 & 0.002 & 0.001 & 0.489 & 0.256 & 0.128 & 0.064 & 0.032 & 0.016 \\
0.016 & 0.008 & 0.004 & 0.002 & 0.001 & 0.489 & 0.256 & 0.128 & 0.064 & 0.032 \\
0.032 & 0.016 & 0.008 & 0.004 & 0.002 & 0.001 & 0.489 & 0.256 & 0.128 & 0.064 \\
0.064 & 0.032 & 0.016 & 0.008 & 0.004 & 0.002 & 0.001 & 0.489 & 0.256 & 0.128 \\
0.128 & 0.064 & 0.032 & 0.016 & 0.008 & 0.004 & 0.002 & 0.001 & 0.489 & 0.256 \\
0.256 & 0.128 & 0.064 & 0.032 & 0.016 & 0.008 & 0.004 & 0.002 & 0.001 & 0.489
\end{pmatrix}.
$$

The cache at each transmitter includes two states: existence and nonexistence of
the requested content. The cache state transition probability matrix is set to be

$$
P_{cache} = \begin{pmatrix}
0.6 & 0.4 \\
0.4 & 0.6
\end{pmatrix}.
$$

Table 3.1 Parameter values used in the simulations

Parameter	Value	Description
Mini-batch size	8	How many experience cases are used for each training step
Update frequency	4	The frequency to perform a training step
Experience replay buffer size	50,000	Training cases are randomly sampled from this number of the most recently experiences
Pre-training steps	10,000	How many steps of random actions are executed before the leaning begins, and the resulting experience are stored to populate the experience replay buffer
Total training steps	500,000	How many steps are used to train the network model
Discount factor	0.99	Discount factor used on the Q-function
Learning rate	0.0001	The learning rate used by AdamOptimizer
Initial exploration	1	The starting chance of random action in the ε-greedy exploration
Final exploration	0.1	The final chance of random action in the ε-greedy exploration
Anneling steps	10,000	How many training steps used to reduce ε from its starting value to the final value
Target network update rate	0.001	The rate to update target Q-network towards primary Q-network

The detailed parameters in the proposed deep reinforcement learning algorithm are listed in Table 3.1. The visualization of the deep Q network model is presented in Fig. 3.2 using TensorBoard, which is a build-in module of TensorFlow. From the graph, it can be clearly visualized that double deep Q-networks are utilized, four convolutional layers are adopted in each deep Q-network, and the advantage and value functions are separately computed. The network models are saved and can be loaded to further train or test the models. In the simulations, some parameters are changed to study the effects of these parameters.

3.4.3 Simulation Results and Discussions

Figure 3.3 shows the convergence performance of the proposed scheme with different learning rates in the deep reinforcement learning algorithm. As we can see from the expression of the system state, the number of the possible states is $2^L \times H^{L^2}$, where 2^L is for the cache status and H^{L^2} is for the status of L^2 equivalent channel coefficients $|h^{[kj]}(t)|^2$ with $k, j = 1, 2, \ldots, L$. In addition to the number of possible states, the complexity of the deep Q learning algorithm depends on many other factors, including the number of actions, the state transition probability, and the rewards. Moreover, since deep Q learning utilizes deep learning to approximate the Q function, the convergence time is affected by many elements, such as the number of convolutional layers, the learning rate, the batch size, and some other parameters in the training process. From Fig. 3.3, we can observe that the average sum rate of the proposed scheme is very low at the beginning of the learning process. With the increase of the number of the episode, the average sum rate increases until it reaches a relatively stable value, which is around 300 Mbps in Fig. 3.3. This shows the convergence performance of the proposed scheme. We can also observe that the learning rate in the AdamOptimizer has effects on the convergence performance in Fig. 3.3. Specifically, the convergence is faster when the learning rate is 0.0001 compared to the case when the learning rate is 0.00001. However, a larger learning rate will result in local optimum instead of global optimum. Therefore, an appropriate learning rate should be chosen for a specific problem. In the rest of the simulations, we choose the learning rate of 0.0001.

Figure 3.4 shows the effects of the mini-batch size for each gradient update in the deep reinforcement learning algorithm on the convergence performance. The parameter of mini-batch size determines how many experience cases are used for each training step. We can see from Fig. 3.4 that the convergence is faster when the mini-batch size is 8 compared to the cases when the mini-batch size is 32 and 64. Similar to the learning rate parameter, an appropriate mini-batch size should be chosen for a specific problem. In the rest of the simulations, we choose the mini-batch size of 8.

Figure 3.5 shows the network's average sum rate with different state-transition probabilities of staying in the same state. Our proposed schemes with (w.) and without (w.o.) cache are compared with an existing selection scheme w. cache [12] and an existing selection scheme w.o. cache [32], where invariant channels are

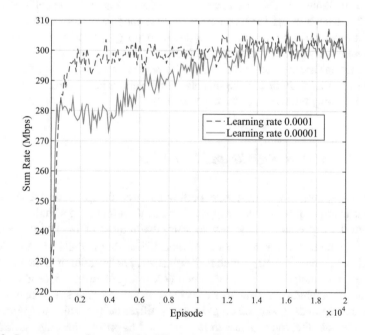

Fig. 3.3 Convergence performance with different learning rates

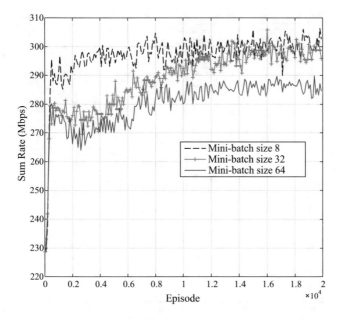

Fig. 3.4 Convergence performance with different batch sizes

assumed for both existing schemes. It can be seen that the proposed OIA w. cache scheme can achieve the highest sum rate compared to the other three schemes. This is because the channel is time-varying, and the proposed scheme can obtain the optimal IA user selection policy in the realistic time-varying channel environment using the deep reinforcement learning algorithm. For both the cases w. cache and w.o. cache, we can observe that the performance of the existing selection method is getting closer to the proposed OIA as the transition probability increases, and this method performs the same when the channel remains absolutely static, i.e., the transition probability that the channel will be in the same state is 1.

Figure 3.6 shows the network's average sum rate with different average SNR values, where the SNR is defined as $10\log_{10}(P^{[k]}/\sigma^2)$ dB, and the same SNR definition applies to other figures in this chapter. From this figure, we can observe that the average sum rate of the proposed OIA scheme w. cache can outperform the other four schemes with different average SNR values. This is because the existing selection scheme assumes time-invariant channels, and the estimated CSI of the current time instant is simply taken as the predicted CSI of the next time instant. In addition, the proposed OIA scheme w.o. cache does not take advantages of caching to relieve the traffic loads of backhaul links, thus has worse performance compared to the proposed scheme w. cache. Note that the existing user selection scheme w. cache performs similarly with the existing scheme w.o. user selection due to the fact that they both considers caching and aims at maximizing the network sum rate, however, one exploits power allocation strategy, and the other one utilizes user selection strategy.

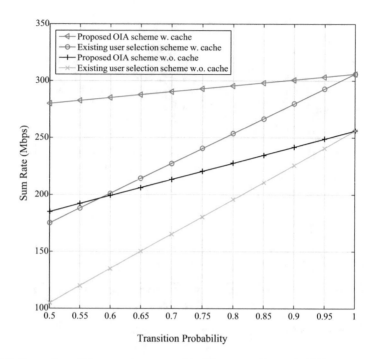

Fig. 3.5 Comparison of the average sum rate with different transition probabilities

Figure 3.7 shows the effects of different ρ values. The average SNR is 30 dB. As defined in (3.12), ρ is the normalized autocorrelation function of a fading channel with motion at a constant velocity. From Fig. 3.7, we can see that the network's average sum rate increases with the increase of ρ in different schemes. This is because a higher ρ value means more accurate CSI, which will result in higher average sum rate in the network. In comparison, the proposed OIA with cache scheme has better performance compared to two other schemes, because the proposed scheme considers the time-varying channels and takes advantage of caching. Figure 3.8 shows the effects of different ρ values when the average SNR is 10 dB. We observe the same trend in Fig. 3.8 with lower average sum rate due to the low SNR value compared to Fig. 3.7.

Figure 3.9 shows the effects of different values of the total backhaul capacity. The average SNR is 30 dB. From Fig. 3.9, we can see that the network's average sum rate increases with the increase of total backhaul capacity in different schemes. This is because a higher value of the total backhaul capacity will provide more capacity for CSI exchange among the transmitters in interference alignment wireless networks, which will result in higher average sum rate in the network. Nevertheless, the effect of the total backhaul capacity is not very significant in the proposed scheme w. cache, because the proposed scheme takes advantages of caching, and the more backhaul capacity is saved for CSI exchange. Figure 3.10 shows the effects of

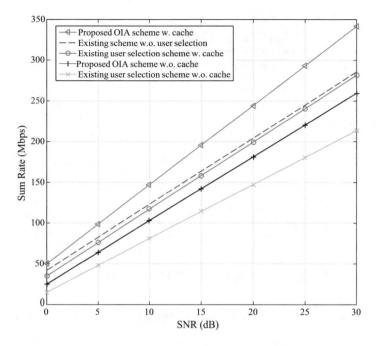

Fig. 3.6 Comparison of the average sum rate with different SNR values

different values of the total backhaul capacity when the average SNR is 10 dB. We can see the same trend in Fig. 3.10 with lower average sum rate due to the low SNR value compared to Fig. 3.9.

Figure 3.11 shows the energy efficiency of proposed scheme and other compared schemes. Except the existing w.o. user selection scheme [33] that exploits power allocation, all the other four schemes are of equal power allocation. The power consumption of circuit blocks is set to be 210 mW. The transmitted power of one user is 100 mW, where for the power allocation scheme, the total power of the network is 500 mW for a 5-user IA network. The energy efficiency of the power amplifier is set to 90%. From this figure, we can observe that the existing scheme w.o. user selection that uses power allocation technique performs better than other existing schemes due to effective power allocation. Nevertheless, it still performs worse than the proposed scheme due to the underlying assumption of time-invariant channels used in the existing scheme w.o. user selection that uses power allocation technique.

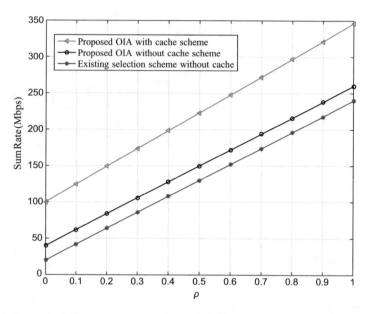

Fig. 3.7 Comparison of the average sum rate with different ρ values (average SNR = 30 dB)

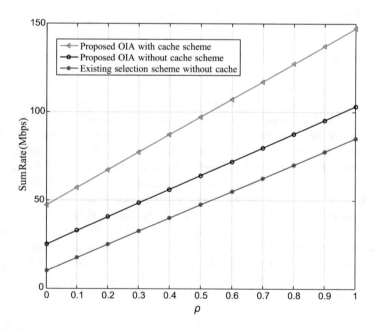

Fig. 3.8 Comparison of the average sum rate with different ρ values (average SNR = 10 dB)

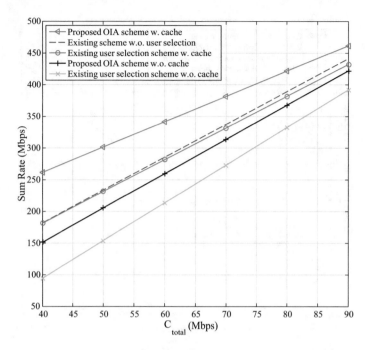

Fig. 3.9 Comparison of the average sum rate with different values of the total backhaul capacity (average SNR = 30 dB)

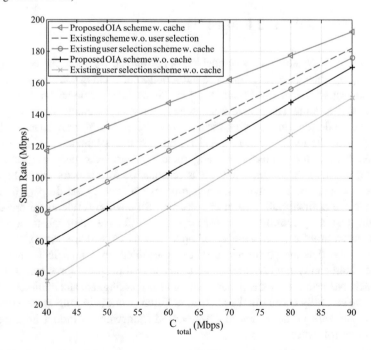

Fig. 3.10 Comparison of the average sum rate with different values of the total backhaul capacity (average SNR = 10 dB)

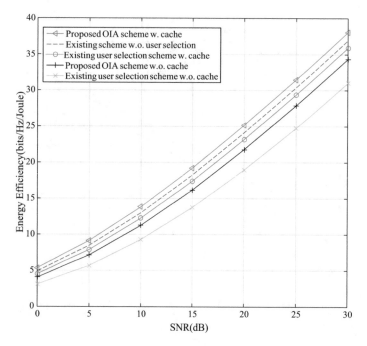

Fig. 3.11 Energy efficiency comparison with different SNR values

3.5   Conclusions and Future Work

In this chapter, we studied cache-enabled opportunistic IA under the condition of time-varying channel coefficients. The introduction of caching can save the limited backhaul capacity that can be utilized for CSI exchange among the transmitters in interference alignment wireless networks. The system complexity is very high when we model the time-varying channel as a finite-state Markov channel. Thus, we exploited the recent advances, and formulated the optimization of the cache-enabled opportunistic interference alignment network as a deep reinforcement learning problem. A central scheduler is responsible for collecting the CSI from each candidate, and then sends the integral system state to the deep Q network to derive the optimal policy for user selection. Simulation results were presented to show that deep reinforcement learning is an effective approach to solve the optimization problem in cache-enabled opportunistic IA wireless networks. It was demonstrated that the performance of cache-enabled opportunistic IA networks can be significantly improved by using the proposed deep reinforcement learning approach. Nevertheless, some parameters, such as learning rate and mini-batch size, should be carefully chosen in the algorithm. Future work is in progress to consider wireless network virtualization in the proposed framework to further improve the network performance.

References

1. G. Paschos, E. Bastug, I. Land, G. Caire, and M. Debbah, "Wireless caching: technical misconceptions and business barriers," *IEEE Comm. Mag.*, vol. 54, no. 8, pp. 16–22, Aug. 2016.
2. C. Liang, F. R. Yu, and X. Zhang, "Information-centric network function virtualization over 5G mobile wireless networks," *IEEE Network*, vol. 29, no. 3, pp. 68–74, May 2015.
3. X. Wang, M. Chen, T. Taleb, A. Ksentini, and V. C. M. Leung, "Cache in the air: exploiting content caching and delivery techniques for 5G systems," *IEEE Commun. Mag.*, vol. 52, no. 2, pp. 131–139, Feb. 2014.
4. C. Fang, F. R. Yu, T. Huang, J. Liu, and Y. Liu, "A survey of green information-centric networking: Research issues and challenges," *IEEE Comm. Surveys Tutorials*, vol. 17, no. 3, pp. 1455–1472, Thirdquarter 2015.
5. D. Liu, B. Chen, C. Yang, and A. F. Molisch, "Caching at the wireless edge: design aspects, challenges, and future directions," *IEEE Commun. Mag.*, vol. 54, no. 9, pp. 22–28, 2016.
6. V. R. Cadambe and S. A. Jafar, "Interference alignment and degrees of freedom of the K-user interference channel," *IEEE Trans. Inform. Theory*, vol. 54, no. 8, pp. 3425–3441, Aug. 2008.
7. C. Suh and D. Tse, "Interference alignment for cellular networks," in *Proc. 46th Annual Allerton Conf. on Commun., Control, and Computing,*. Monticello, IL, Sep. 2008, pp. 1037–1044.
8. S. M. Perlaza, N. Fawaz, S. Lasaulce, and M. Debbah, "From spectrum pooling to space pooling: opportunistic interference alignment in MIMO cognitive networks," *IEEE Trans. Signal Proc.*, vol. 58, no. 7, pp. 3728–3741, 2010.
9. X. Li, N. Zhao, Y. Sun, and F. R. Yu, "Interference alignment based on antenna selection with imperfect channel state information in cognitive radio networks," *IEEE Trans. Veh. Tech.*, vol. 65, no. 7, pp. 5497–5511, July 2016.
10. B. C. Jung and W.-Y. Shin, "Opportunistic interference alignment for interference-limited cellular TDD uplink," *IEEE Commun. Lett.*, vol. 15, no. 2, pp. 148–150, 2011.
11. Y. He, H. Yin, and N. Zhao, "Multiuser-diversity-based interference alignment in cognitive radio networks," *AEU-Int. J. Electron. C*, vol. 70, no. 5, pp. 617–628, 2016.
12. M. Deghel, E. Baştuğ, M. Assaad, and M. Debbah, "On the benefits of edge caching for MIMO interference alignment," in *Proc. IEEE SPAWC*, 2015, pp. 655–659.
13. M. A. Maddah-Ali and U. Niesen, "Cache-aided interference channels," in *Proc. IEEE ISIT*, 2015, pp. 809–813.
14. O. E. Ayach, S. W. Peters, and R. W. Heath, "The practical challenges of interference alignment," *IEEE Wirel. Commun.*, vol. 20, no. 1, pp. 35–42, Feb. 2013.
15. J. Yang, A. K. Khandani, and N. Tin, "Statistical decision making in adaptive modulation and coding for 3G wireless systems," *IEEE Trans. Veh. Technol.*, vol. 54, no. 6, pp. 2066–2073, 2005.
16. A. Z. Ghanavati, U. Pareek, S. Muhaidat, and D. Lee, "On the performance of imperfect channel estimation for vehicular ad-hoc networks," in *Proc. IEEE VTC'10Fall*, Sept. 2010, pp. 1–5.
17. R. Xie, F. R. Yu, and H. Ji, "Dynamic resource allocation for heterogeneous services in cognitive radio networks with imperfect channel sensing," *IEEE Trans. Veh. Tech.*, vol. 61, pp. 770–780, Feb. 2012.
18. Y. Cai, F. R. Yu, C. Liang, B. Sun, and Q. Yan, "Software defined device-to-device (D2D) communications in virtual wireless networks with imperfect network state information (NSI)," *IEEE Trans. Veh. Tech.*, no. 9, pp. 7349–7360, Sept. 2016.
19. Y. He, F. R. Yu, N. Zhao, H. Yin, H. Yao, and R. C. Qiu, "Big data analytics in mobile cellular networks," *IEEE Access*, vol. 4, pp. 1985–1996, Mar. 2016.
20. Y. Wei, F. R. Yu, and M. Song, "Distributed optimal relay selection in wireless cooperative networks with finite-state Markov channels," *IEEE Trans. Veh. Technol.*, vol. 59, no. 5, pp. 2149–2158, 2010.

21. H. S. Wang and P.-C. Chang, "On verifying the first-order Markovian assumption for a Rayleigh fading channel model," *IEEE Trans. Veh. Technol.*, vol. 45, no. 2, pp. 353–357, 1996.
22. C. Luo, F. R. Yu, H. Ji, and V. C. M. Leung, "Cross-layer design for TCP performance improvement in cognitive radio networks," *IEEE Trans. Veh. Tech.*, vol. 59, no. 5, pp. 2485–2495, June 2010.
23. V. Mnih, K. Kavukcuoglu, D. Silver, A. A. Rusu, J. Veness, M. G. Bellemare, A. Graves, M. Riedmiller, A. K. Fidjeland, G. Ostrovski *et al.*, "Human-level control through deep reinforcement learning," *Nature*, vol. 518, no. 7540, pp. 529–533, 2015.
24. Y. He, C. Liang, F. R. Yu, N. Zhao, and H. Yin, "Optimization of cache-enabled opportunistic interference alignment wireless networks: A big data deep reinforcement learning approach," in *Proc. IEEE ICC'17*, Paris, France, June 2017.
25. Y. He, F. R. Yu, N. Zhao, V. C. M. Leung, and H. Yin, "Software-defined networks with mobile edge computing and caching for smart cities: A big data deep reinforcement learning approach," *IEEE Commun. Mag.*, vol. 55, no. 12, Dec. 2017.
26. K. Gomadam, V. R. Cadambe, and S. A. Jafar, "A distributed numerical approach to interference alignment and applications to wireless interference networks," *IEEE Trans. Inform. Theory*, vol. 57, no. 6, pp. 3309–3322, 2011.
27. A. Tatar, M. D. de Amorim, S. Fdida, and P. Antoniadis, "A survey on predicting the popularity of web content," *Springer J. Internet Services and Applications*, vol. 5, no. 1, p. 8, 2014.
28. D. Tse and P. Viswanath, *Fundamentals of Wireless Communication*. Cambridge, U.K.: Cambridge Univ. Press, 2005.
29. R. H. Clarke, "A statistical theory of mobile radio reception," *Bell Syst. Tech. J.*, vol. 47, no. 6, pp. 957–1000, Jul.-Aug. 1968.
30. M. Abadi, A. Agarwal *et al.*, "Tensorflow: Large-scale machine learning on heterogeneous systems," *arXiv:1603.04467*, Nov. 2015.
31. "Tensorflow.org," https://www.tensorflow.org/.
32. N. Zhao, F. R. Yu, H. Sun, and M. Li, "Adaptive power allocation schemes for spectrum sharing in interference alignment (IA)-based cognitive radio networks," *IEEE Trans. Veh. Tech.*, vol. 65, no. 5, pp. 3700–3714, May 2016.
33. F. Cheng, Y. Yu, Z. Zhao, N. Zhao, Y. Chen, and H. Lin, "Power allocation for cache-aided small-cell networks with limited backhaul," *IEEE Access*, vol. 5, pp. 1272–1283, 2017.
34. C. M. Yetis, T. Gou, S. A. Jafar, and A. H. Kayran, "On feasibility of interference alignment in MIMO interference networks," *IEEE Trans. Signal Proc.*, vol. 58, no. 9, pp. 4771–4782, 2010.
35. H. S. Wang and N. Moayeri, "Finite-state Markov channel-a useful model for radio communication channels," *IEEE Trans. Veh. Tech.*, vol. 44, no. 1, pp. 163–171, Feb. 1995.

Chapter 4
Deep Reinforcement Learning for Mobile Social Networks

Abstract Social networks have continuously been expanding and trying to be innovative. The recent advances of computing, caching, and communication (3C) can have significant impacts on mobile social networks (MSNs). MSNs can leverage these new paradigms to provide a new mechanism for users to share resources (e.g., information, computation-based services). In this chapter, we exploit the intrinsic nature of social networks, i.e., the trust formed through social relationships among users, to enable users to share resources under the framework of 3C. Specifically, we consider the mobile edge computing (MEC), in-network caching and device-to-device (D2D) communications. When considering the trust-based MSNs with MEC, caching and D2D, we apply a novel deep reinforcement learning approach to automatically make a decision for optimally allocating the network resources. The decision is made purely through observing the network's states, rather than any handcrafted or explicit control rules, which makes it adaptive to variable network conditions. Google TensorFlow is used to implement the proposed deep Q-learning approach. Simulation results with different network parameters are presented to show the effectiveness of the proposed scheme.

4.1 Introduction

Mobile social networks (MSNs) have been developing rapidly, which can provide a variety of social services and applications to mobile users [1]. Millions of mobile users interact with each other in MSNs, and MSNs will become one of the most important networking paradigms in future wireless mobile networks [2]. MSNs have always been trying to be innovative and leverage new technologies. The recently proposed integrated framework of computing, caching and communication (3C) can have positive impacts on MSNs. The integration of the 3C framework and MSNs can help create a new mechanism to share resources among users. The available resources that can be shared extend beyond information, for instance, the storage devices and computing capability contributed by users can also be shared. Additionally, the MSNs have an inherent advantage, i.e., the trust of mobile users is naturally formed through the social relationships and interactions. This makes

© The Author(s), under exclusive license to Springer Nature Switzerland AG 2019   45
F. R. Yu, Y. He, *Deep Reinforcement Learning for Wireless Networks*, SpringerBriefs
in Electrical and Computer Engineering, https://doi.org/10.1007/978-3-030-10546-4_4

the resource sharing between users more reliable. In this chapter, we consider the specific 3C framework with *mobile edge computing* (MEC), *in-network caching*, and *device-to-device* (D2D) communications.

With MEC, computation resources are placed at the edge of wireless mobile networks in physical proximity to mobile users [3]. Compared to traditional mobile cloud computing, MEC can provide faster interactive response by low-latency connections. Therefore, MEC has been envisioned as a promising technique to offer agile and ubiquitous computation augmenting mobile services and applications, including social services and applications [4, 5]. Another promising technology is *in-network caching*, which can effectively reduce the duplicate content transmission in the network. Recent studies of applying the in-network caching technique in MSNs show that traffic loads and latency can be significantly reduced by caching contents in MSNs [2]. In addition, D2D communications can be beneficial to MSNs as well [6]. With D2D communications, users in close proximity can directly communicate with each other via D2D links, instead of accessing base stations (BSs) exclusively. When it comes to the content-centric MSNs, in spite of the smaller-sized storage (compared to that of BSs), the ubiquitous caching capability residing in mobile devices cannot be neglected due to their ubiquitous in-network distribution and ever-increasing storage size [6, 7].

Although some works have been done on applying recent advances of MEC, in-network caching and D2D to improve the performance of MSNs, the *knowledge of social relationships* among users in MSNs is largely ignored in these new paradigms to improve the reliability and efficiency of resource sharing and delivery in MSNs. In fact, social relationships have been investigated to enhance wireless networking. For example, Zhang et al. [8] exploit social ties to achieve a reliable D2D communication, which can improve packet transmission and reduce the workload on the network infrastructures. Pan et al. [9] utilize social information as an essential element for designing forwarding algorithms in ad hoc networks. Social relationships also play an important role for routing in wireless environments [10].

Furthermore, the dynamic nature of the available resources has not been paid much attention in the existing literature. To fill in this gap, in this chapter, we study trust-based social networks with recent advances of MEC, caching and D2D. Considering the integrated network, the allocation of resources for subscribed users is complicated, especially when the conditions of the network resources are varying with time. Therefore, we utilized a novel deep reinforcement learning approach to automatically achieve the resource allocation tasks.

In the following, we first review some of the existing excellent works that focus on efficient management and allocation schemes of computing, caching and communication resources. Some of the proposed schemes only consider one of these three types of resources. On the other hand, in some other works, two or three types of resources are jointly considered to effectively and efficiently achieve optimal performance metrics. Then, we discuss our contributions in this chapter.

4.1.1 Related Works

With the increasing popularity of compute intensive applications, such as augmented reality, interactive video games, etc., MEC is becoming a very promising paradigm that enables the possibility of offloading the computation tasks from resource-limited mobile devices to more powerful edge servers. MEC can bring various benefits, where minimizing energy consumption and minimizing latency are two major optimization objectives [11]. In the literature, various resource allocation approaches have been proposed to solve the formulated optimization problems. For example, Zhang et al. [12] propose an energy-efficient computation offloading scheme that jointly optimizes radio resource allocation and offloading to minimize the energy consumption of the offloading system with the latency constraint. In their scheme, the mobile devices are first classified into three types and then wireless channels are allocated to mobile devices based on their priority iteratively. By doing this, the optimization problem can be solved in polynomial complexity. Liu et al. [13] design an optimal computation task scheduling policy for MEC systems. They first analyze the average delay of each task and average power consumption at the mobile device side under the proposed scheduling policy using Markov chain theory. Then they formulate a delay minimization problem with the power constraint. An efficient one-dimensional search algorithm is used to derive the optimal offloading policy. In [14], the authors jointly optimize the radio resources and computational resources to minimize the total energy expenditure, where an iterative algorithm based on successive convex approximation technique is adopted to solve the formulated non-convex optimization problem. In [15], a distributed game theoretic approach is proposed to solve the NP-hard efficient computation offloading problem. Chen et al. [16] choose a model-free RL technique to solve the optimal traffic offloading strategy for heterogeneous cellular networks with dynamic traffic.

When caching technology is integrated into mobile networks, where to cache, what to cache, and how to cache are the major factors that make a significant influence on the system performance [11]. In [17], the authors investigate the proactive storage allocation problem over BSs in cellular networks, which is proved to be NP-hard. To get a low-complexity solution, a heuristic method is utilized and the convergence is proved by strict theoretic deduction. A caching-enabled D2D communication scheme is designed in [18], where the caching strategy is optimized by jointly considering users' social relationships and common interests with the constraint of hit ratio, delay and caching capacity. In fact, contents with the most popularity should be cached within limited caching capacity. Since the content popularity is varying with time, learning based caching policies are proposed in [19]. The authors formulate the distributed caching over small BSs (SBSs) as a reinforcement learning problem, and with the help of coded caching, the complicated caching problem is solved by being reduced to a convex optimization problem. In [20], Qiao et al. consider the mobility pattern of mobile users, and formulate the mobility-aware caching problem as an optimization problem to maximize the caching utility.

The problem is solved by using a polynomial-time heuristic solution. He et al. exploit deep reinforcement learning to address the resource allocation problems in cache-enabled interference alignment networks [21, 22].

The comprehensive resource allocation schemes for efficient integration of computing, caching and communication resources to achieve the optimal performance have been developed, even though not yet been widely investigated. In [23], Zhou et al. design a novel information-centric heterogeneous network framework, and with the virtualization technology, communication, computing and caching resources are shared among users. They formulate the virtual resource allocation strategy as a joint optimization problem, which is a non-convex NP hard problem. An alternating direction method of multiplier (ADMM) approach is used to solve the problem by simplifying and relaxing the non-convex problem into a convex one. ADMM method is also used in [24] to achieve the optimal resource allocation strategy in wireless networks with MEC and in-network caching being jointly considered. He et al. propose an integrated framework that can enable dynamic orchestration of networking, caching and computing resources, where the dynamics are modeled as finite-state Markov chains [25, 26]. The authors use the deep Q-leaning approach to solve the complex problems with large numbers of system states and actions. Research on this topic of resource allocation strategies for integrated systems of computing, caching and communication will continue with the evolving of each of these three technologies.

4.1.2 Contributions

The main contributions of this chapter are summarized as follows.

- In this chapter, we consider trust-based social networks specifically with MEC, in-network caching and D2D communications under the umbrella of a 3C framework. An optimization problem is formulated to maximize the network operator's utility with comprehensive considerations of trust values, computation capabilities, wireless channel qualities, and the cache status of all the available BSs and D2D nodes.
- To be more realistic, we consider that the network conditions (i.e., trust value, computation capability, wireless channels, and cache status) are varying with time, and the dynamics of the computing capabilities, cache status and wireless channel conditions are modeled as Markov chains. In the meanwhile, the trust values are derived from both direct and indirect observations using Bayesian inference and Dempster-Shafer theory, respectively.
- The complexity of the integrated network is very high when we jointly consider the dynamic trust values, computation capabilities, wireless channel conditions and the cache status, and it is extremely difficult to solve the formulated optimization problem. In this chapter, we exploit deep Q-learning based resource allocation strategy to solve the optimization problem without any explicit assumptions or simplifications.

- Google TensorFlow is used to implement the proposed deep Q-learning approach. Simulation results with different system parameters are presented to show the effectiveness of the proposed scheme. It is illustrated that the performance of MSNs can be significantly improved with the proposed approach.

The remainder of this chapter is outlined as follows. We describe the system model in Sect. 4.2, which includes system description, network model, communication model, caching model and computing model. Next, the social trust scheme with uncertain reasoning is presented in Sect. 4.3. Section 4.4 formulates this system as a deep reinforcement learning problem. Simulation results are presented and discussed in Sect. 4.5. Finally, we give the conclusions and future work in Sect. 4.6.

4.2 System Model

In this section, we first present the system description. Next, the network model, communication model, cache model and computing model are presented, respectively.

4.2.1 System Description

MSNs have been developed rapidly to provide a variety of social services and applications to mobile users, focusing not only on the behaviors but also on the social needs of the users [1]. Compared to conventional mobile wireless networks, mobile users in MSNs do not always contact remote servers to request contents. Instead, mobile users in MSNs can directly obtain contents from each other within a community based on their social ties [8, 27].

In MSNs, huge amounts of information-rich data will be exchanged by mobile users. Although cloud computing is powerful, it is difficult for data centers to provide mobile users with low-latency services. To address these issues, MEC has been proposed to deploy computing resources closer to end users, which can efficiently improve the quality of service (QoS) for applications that require intensive computations and low latency [4, 28]. MEC applies the concept of cloud computing in network edge nodes, to facilitate services and applications, including mobile social services and applications.

In addition, due to user mobility and poor-quality wireless radio links, it is challenging to deliver huge amounts of data using the traditional client-server approach in MSNs [2]. Recent advances of in-network caching can be extended to MSNs to address this issue, which can efficiently reduce the duplicate content transmission in networks. This innovative content-centric approach has been studied in MSNs, which natively privilege the information (e.g., trusted information in a specific proximity of an event and a specific time period) rather than the node identity. In addition, with in-network caching, mobility and sporadic connectivity issues can be effectively addressed in MSNs.

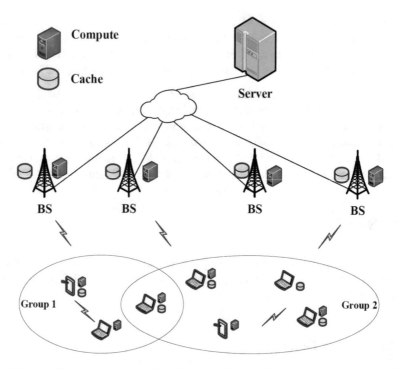

Fig. 4.1 A mobile social network with edge computing, caching and device-to-device (D2D) communications

Moreover, with D2D communications, users in close proximity can directly communicate with each other via D2D links, instead of accessing BSs exclusively. As a promising approach to offload traffic from BSs, D2D communications can enable the sharing of radio connectivity and direct information delivery between two close users [29, 30].

Security is always an important aspect in wireless applications and services [31, 32]. Trust-based security schemes are important detection-based approaches in MSNs. The definition of trust in MSNs is similar to that in sociology, where trust is interpreted as degrees of the belief that an entity will carry out tasks as expected [33]. In this chapter, we present a framework for trust-based social networks specifically with MEC, in-network caching and D2D communications, which is depicted in Fig. 4.1. Under the framework, we illustrate a video content request task as an example, which will be elaborated in the following network model.

4.2.2 Network Model

In this chapter, we consider the scenario with K BSs, and M mobile users that are considered as transmitters in the potential D2D communications. The network is operated by a central controller. Assume that there are L subscribed mobile users issuing video requests to the network operator, which is responsible for assigning a proper content provider to each requester, i.e., associating with one BS or setting up a one-to-one D2D communication. The sets are denoted as $\mathscr{C}_K = \{1, 2, \ldots, K\}$, $\mathscr{C}_M = \{1, 2, \ldots, M\}$ and $\mathscr{L} = \{1, 2, \ldots, L\}$, respectively. All the K BSs, the M D2D transmitters and the L mobile users are equipped with both caching and computing capabilities. The unicast D2D communication is available only when the mobile user's requested video is stored in the transmitter's cache. In addition, the transmitters can also provision mobile edge computing if they have spare computing capacity left at the considered moment.

In our system model, we assume that the lth mobile user, denoted as s_l, issues a video request. First, the network controller checks all the video providers, i.e., all the BSs and D2D transmitters. We denote $\mathscr{C} = \mathscr{C}_K \bigcup \mathscr{C}_M$ be the set of video providers, and let $c_i (1 \leq i \leq K)$ be the ith BS and $c_j (K + 1 \leq j \leq K + M)$ be the jth transmitter. Each video provider's cache is labeled by a content indicator which indicates whether or not the requested video is being stored.

If the requested video is being stored at any of the video provider's caches, and the versions match up as well, the network controller will built up a communication between mobile user s_l and the optimal video provider. However, if all the stored videos' versions mismatch with the requested one, video transcoding should be performed at either mobile user s_l itself or at the video provider's side. On the other hand, If the requested video is not being stored at any of the caches, it has to associate with a proper BS.

Note that in this work, we use video version to represent video specification (e.g., H.263, H.264, MPEG2, or MPEG4). With the rapid growth of mobile services, increasing volumes of videos are played by mobile devices. Consequently, service providers often need to transcode the video contents into different specifications (e.g., bit rate, resolution, quality, etc.) with different QoS (e.g., delay) for heterogeneous mobile devices, networks, and user preferences.

4.2.3 Communication Model

Denote $h_{s_l}^{c_p}$ as the channel gain between mobile user s_l and content provider c_p, $\forall l \in \mathscr{L}, \forall p \in \mathscr{C}$. According to the Shannon theorem, the achievable rate of mobile user s_l when associating with provider c_p can be expressed as

$$r_{s_l}^{c_p} = B \log_2 \left(1 + \frac{p_T h_{s_l}^{c_p}}{N_o B} \right), \tag{4.1}$$

where B denotes the bandwidth allocated to each mobile user, p_T is the equally transmitted power, and N_o is the noise spectral density.

Here, we consider the realistic wireless channels, and actually $h_{s_l}^{c_p}$ is a continuous random variable. Such assumption is intractable for analysis [34], and therefore we model the wireless channels as the finite-state Markov channels (FSMC), which may achieve performance improvement compared with the traditional assumption of static channels. In our model, the channel gain $h_{s_l}^{c_p}$ is discretized and quantized into H levels: $\mathcal{H}_0$, if $h_0^* \leqslant h_{s_l}^{c_p} < h_1^*$; $\mathcal{H}_1$, if $h_1^* \leqslant h_{s_l}^{c_p} < h_2^*$; $\ldots$; $\mathcal{H}_{H-1}$, if $h_{s_l}^{c_p} \geqslant h_{H-1}^*$. The boundary values for a particular environment under certain criterion should be optimized for better performance [35]. However, for simplicity, uniformly setting the boundary values of FSMC is widely used in the literature (e.g., [35]). In this chapter, all the boundary values from h_1^* to h_{H-1}^* are increasing in the same distance with each other. Each level corresponds to a state of the Markov chain, and therefore a H-element state space is formed. Due to the relationship between $h_{s_l}^{c_p}$ and $r_{s_l}^{c_p}$, for convenience we use $r_{s_l}^{c_p}$ to describe a state of the wireless channel. We consider the dynamic process, and the channel state realization of $r_{s_l}^{c_p}$ at time instant t is denoted as $\Upsilon_{s_l}^{c_p}(t)$. Based on a certain transitional probability, $\Upsilon_{s_l}^{c_p}(t)$ varies from one state to another as one time slot is gone. The transitional probability of $\Upsilon_{s_l}^{c_p}(t)$ from one state j_s to another state k_s is denoted as $\psi_{j_s k_s}(t)$. The $H \times H$ channel state transitional probability matrix between mobile user s_l and provider c_p is shown as:

$$\Psi_{s_l}^{c_p}(t) = \left[\psi_{j_s k_s}(t) \right]_{H \times H}, \tag{4.2}$$

where $\psi_{j_s k_s}(t) = \Pr \left(\Upsilon_{s_l}^{c_p}(t+1) = k_s \mid \Upsilon_{s_l}^{c_p}(t) = j_s \right)$.

4.2.4 Cache Model

Assume that there are totally I video contents in the network that mobile users can request for. The set of the contents is denoted as $\mathcal{I} = \{1, 2, \ldots, I\}$, which is ranked by popularity. Here, we consider finite cache capacity, i.e., each provider caches some of the I video contents, and refreshes the contents periodically. We consider that mobile user s_l has a video request $v_i, i \in \mathcal{I}$. After receiving the request message, the network controller checks every provider's content indicator for video v_i, i.e., $x_{c_p}^{v_i}, \forall p \in \mathcal{C}$. The content distribution indicator $x_{c_p}^{v_i} = 1$ means that video v_i is being stored in the cache of provider p; otherwise $x_{c_p}^{v_i} = 0$. Here, $x_{c_p}^{v_i}$ is viewed as a random variable, representing the state of the cache, and is modeled using a two-state (i.e., state 0 and 1) Markov chain [36]. The transitional probability matrix of the state $x_{c_p}^{v_i}$ is defined as:

$$\Gamma_{c_p}^{v_i}(t) = [\delta_{a_s b_s}(t)]_{2 \times 2}, \tag{4.3}$$

where $\delta_{a_s b_s}(t) = \Pr \left(x_{c_p}^{v_i}(t+1) = b_s \mid x_{c_p}^{v_i}(t) = a_s \right)$, and $a_s, b_s \in \{0, 1\}$.

When the cache capacity is assumed to be finite, the transitional probability matrix of the cache state can be derived based on different cache refreshment strategies. An important one is the least recently used (LRU) cache refreshment policy. The transitional probability matrix can be obtained using the following

Markov chain flow matrix [36],

$$
\Lambda_i =
\begin{bmatrix}
-\gamma_i & 0 & \cdots & 0 & 0 & \gamma_i \\
\zeta_i + \mu_i & -\beta - \mu_i & \cdots & 0 & 0 & \gamma_i \\
\vdots & \vdots & \vdots & \vdots & \vdots & \vdots \\
\mu_i & 0 & \vdots & \zeta_i & -\beta - \mu_i & \gamma_i \\
\mu_i & 0 & \vdots & 0 & \zeta_i & -\zeta_i - \mu_i
\end{bmatrix}.
\tag{4.4}
$$

Here $\zeta_i = \beta - \gamma_i$, and γ_i represents the average request rate for the ith popular video v_i that can be denoted as

$$
\gamma_i(t) = \frac{\beta}{\rho i^\alpha},
\tag{4.5}
$$

where the request for v_i is assumed to arrive as a Poisson process with rate β [36], and the probability follows a Zipf-like distribution. Therefore, the probability of requesting content v_i is $1/\rho i^\alpha$, where $\rho = \sum_{i=1}^{I} 1/i^\alpha$ and α is the Zipf slope with $0 < \alpha < 1$. The video content is refreshed periodically, and the lifetime of any video content v_i follows an exponential distribution with mean $1/\mu_i$.

4.2.5 Computing Model

On the condition that the versions of the requested video at all the providers' caches do not match with the requested version, the associated provider has to extract the current video content and construct a computation task based on the input data information, as well as the number of CPU cycles. In the computation model, we construct the computation task as $Q_{s_l}^{v_i} = \{o_{s_l}^{v_i}, q_{s_l}^{v_i}\}$. The first parameter $o_{s_l}^{v_i}$ represents the size of the requested-version video, and the second parameter $q_{s_l}^{v_i}$ indicates the required number of CPU cycles that is needed to finish the computation task.

The computation capability of provider c_p assigned to mobile user s_l is denoted as $f_{s_l}^{c_p}$, which can be measured by the number of CPU cycles per second [15, 23]. In our network, multiple mobile users may associate with the same provider and share the computing device simultaneously, which would result in the fact that the computation capability for provider c_p is not exactly known at the next time instant. Therefore, the computation capability $f_{s_l}^{c_p}$ is modelled as a random variable, and equally divided into N discrete levels denoted by $\mathscr{E} = \{\mathscr{E}_0, \mathscr{E}_1, \ldots, \mathscr{E}_{N-1}\}$. The realization of the random variable $f_{s_l}^{c_p}$ is denoted as $F_{s_l}^{c_p}$ at time slot t. We model the transition of the computation capability level of a provider as a Markov process. The transitional probability matrix of $F_{s_l}^{c_p}$ can be expressed as:

$$
\Theta_{s_l}^{c_p}(t) = \left[\iota_{x_s y_s}(t) \right]_{N \times N},
\tag{4.6}
$$

where $\iota_{x_s y_s}(t) = \Pr\left(F_{s_l}^{c_p}(t+1) = y_s \mid F_{s_l}^{c_p}(t) = x_s \right)$, and $x_s, y_s \in \mathscr{E}$.

The consumed time for executing the task $Q_{s_l}^{v_i}$ at provider c_p can be obtained as $t_{s_l,c_p} = \dfrac{q_{s_l}^{v_i}}{F_{s_l}^{c_p}(t)}$. Furthermore, the rate of computation, i.e., the number of bits computed per second, can be given

$$r_{s_l,c_p}^{\text{comp}}(t) = a_{s_l,c_p}^{\text{comp}}(t)\frac{o_{s_l}^{v_i}}{t_{s_l,c_p}} = a_{s_l,c_p}^{\text{comp}}(t)\frac{f_{s_l}^{c_p}(t)o_{s_l}^{v_i}}{q_{s_l}^{c_p}}, \qquad (4.7)$$

where $a_{s_l,c_p}^{\text{comp}}(t) \in \{0, 1\}$ represents whether or not the computation task is decided to be performed at provider c_p.

4.3   Social Trust Scheme with Uncertain Reasoning

In this section, we will derive how to obtain the trust values of mobile users. We evaluate the trustworthiness of a mobile user by a real number Tr ranging from 0 to 1. In our model, the trust value Tr is jointly determined based on direct observations and indirect observations. The direct observation trust of a mobile user is defined as the estimated degree of trustworthiness from its directly-connected mobile users based on their past experiences. However, the subjective evaluation from direct connections may be prejudiced, and therefore in order to be more objective and impartial, we also consider the rating of trust from other indirectly-connected mobile users. Here, we denote the trust value from direct observations as Tr^D and the trust value from indirect observations as Tr^{InD}. By combining these two trust values, we can obtain a more accurately estimated trust value of a mobile user as

$$Tr = \omega Tr^D + (1 - \omega)Tr^{InD} \qquad (4.8)$$

where ω is the weight coefficient to adjust the weightiness between direct and indirect observations, and $0 \leq \omega \leq 1$.

The trust evaluation procedure in our model can be visually explained as Fig. 4.2. In the following, we will discuss how to obtain the trust evaluation from direct

Fig. 4.2 Social trust evaluation with both direction observation and indirect observation

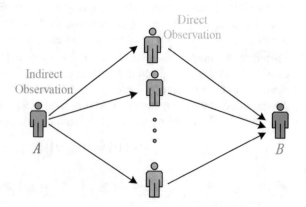

observations and indirect observations by using the Bayesian inference approach and Dempster-Shafer theory [37], respectively.

4.3.1 Trust Evaluation from Direct Observations

In the direct observations, consider that an observing mobile user can overhear the data forwarded by the observed mobile user, and identify the observed mobile user's malicious behaviors, such as discarding or modifying some of the original data.

Through multiple observations of the observed mobile user's behavior, the observing mobile user can evaluate the trust value by exploiting Bayesian inference [38], which is a method of statistical inference using the Bayes' theorem to update the probability for a hypothesis with more evidence becomes available.

Under the Bayesian framework, we model the trust of a mobile user as a continuous random variable, denoted as Φ, where ϕ takes values from 0 to 1. We assume that Φ follows a beta distribution [39], i.e., $\Phi \sim Beta(a, b)$, which is defined as follows with parameters a and b

$$Beta(a, b) = \frac{\phi^{a-1}(1 - \phi)^{b-1}}{\int_0^1 \phi^{a-1}(1 - \phi)^{b-1} \, d\phi}, \tag{4.9}$$

for $0 \leq \phi \leq 1$. Here, since Φ is assumed to obey a beta distribution, the trust value can be represented by the two parameters a and b.

We summarize our belief of the trust Φ in a probability distribution iteratively as more observations are available. Assume that the prior probability density function (pdf) at the $(t - 1)$th observation is known. Then, according to the Bayes theorem, the posterior distribution at the tth observation can be obtained with the pdf as

$$f_t(\phi) = \frac{f_t(x_t|\phi, y_t) f_{t-1}(\phi)}{\int_0^1 f_t(x_t|\phi, y_t) f_{t-1}(\phi) \, d\phi}, \tag{4.10}$$

where x_t and y_t are the number of data packets that has been forwarded correctly and the number of packets received by the observed mobile user at the tth observation, respectively; $f_t(x_t|\phi, y_t)$ is the likelihood function, which follows a binomial distribution as

$$f_t(x_t|\phi, y_t) = \binom{y_t}{x_t} \phi^{x_t} (1 - \phi)^{y_t - x_t}. \tag{4.11}$$

In Bayesian inference, the beta distribution is the conjugate prior probability distribution for the binomial distribution [38, 39]. Since the likelihood function $f_t(x_t|\phi, y_t)$ follows a binomial distribution, the priori distribution $f_{t-1}(\phi)$ is definitely assumed to follow a beta distribution, which reflects what is already known about the distribution of Φ at the $(t - 1)$th observation. Given that the priori distribution $f_{t-1}(\phi)$ follows a beta distribution, the posterior distribution $f_t(\phi)$ also follows a beta distribution. Particularly, if $f_{t-1}(\phi) \sim Beta(a_{t-1}, b_{t-1})$ and x_t, y_t

from the tth observation are also given, then

$$f_t(\phi) \sim Beta(a_{t-1} + x_t, b_{t-1} + y_t - x_t), t \geq 1. \tag{4.12}$$

At the very beginning of the observation, there is no evidence about the distribution of Φ, thus we assume that it follows a uniform distribution, i.e., $f_0(\phi) \sim Beta(1, 1)$. Summarily, we can say that $f_t(\phi)$ follows $Beta(a_t, b_t)$ with parameters

$$a_t = a_{t-1} + x_t, \qquad b_t = b_{t-1} + y_t - x_t,$$
$$a_0 = 1, \qquad\qquad b_0 = 1. \tag{4.13}$$

The trust value can be represented with the mathematical expectation of beta distribution as

$$\mathbb{E}_t[\Phi] = \frac{a_t}{a_t + b_t}. \tag{4.14}$$

Intuitively, the trust value of a mobile user is 0.5 at the beginning, and updated continuously by using the follow-up observations.

From the above discussion, we can see that past experiences play an important role in the Bayesian inference. In fact, recent behaviors of a mobile user should weight more in the trust evaluation. Here, we introduce a punishment factor for reputation fading, which gives more weights on misbehaviors in the Bayesian inference. The trust evaluation formula in (4.14) is revised as follows:

$$\mathbb{E}_t[\Phi] = \frac{a_t}{a_t + \tau b_t}, \tag{4.15}$$

where τ is the punishment factor, and $\tau \geq 1$. With the increment of τ, the trust value declines quickly.

The punishment factor makes the trust evaluation more realistic and reliable. First, if a mobile user once behaves maliciously, compared with those who have no bad records, its trust value will be lowered much more. Second, the trust value will not recover quickly even if he behaves well recently because of the constraint of the punishment factor. This helps distinguish the malicious mobile users quickly and prevent them disrupting others' trust evaluation. Based on the above deduction, the trust value from direct observation, Tr^D, is defined as:

$$Tr^D = \mathbb{E}_t[\Phi]. \tag{4.16}$$

4.3.2 Trust Evaluation from Indirect Observations

Apart from the direct observations, indirect observations from other mobile users are also very important in assessing the trustworthiness of an observed mobile user. Considering the indirect observations helps mitigate the situation that an observed mobile user is loyal to one mobile user but cheating on others. Assume that an observing mobile user collects observations from several other mobile users (also called subsidiary observing mobile user), and combines the collected evidence into a decision about the observed mobile user's trust value. However, these subsidiary observing mobile users may be untrustworthy or the evidence offered by them is unreliable.

The Dempster-Shafer theory can be used as an effective way to handle the uncertainty issue and combine the evidence from multiple subsidiary observers [40]. The core of this theory is based on two ideas: the degrees of belief about a proposition can be obtained from multiple subjective probabilities of a related theme, and these degrees of belief can be combined together under the condition that they are from independent evidence [37]. In the indirect observation, we assume that there are more than one subsidiary observing mobile users and the evidence provided by them is mutually independent.

Belief Function

As an introduction to the belief function, suppose that each subsidiary observing mobile user has two states, i.e., trustworthy and untrustworthy with probabilities p_1 and $1 - p_1$ respectively. Assume that mobile user 1 claims that the observed mobile user B is trustworthy. If mobile user 1 is trustworthy, its statement is regarded as true; however, if mobile user 1 is untrustworthy, its statement is not necessarily untrue. We think that mobile user 1 provides p_1 degree of belief in the observed mobile user's trustworthiness (hypothesis H), 0 degree of belief in the untrustworthiness (hypothesis $\overline{H}$), and $1 - p_1$ degree in the uncertainty, i.e., either trustworthy or untrustworthy (hypothesis U). Each hypothesis is assigned with a basic probability value $m(H)$, $m(\overline{H})$, $m(U)$ taking values from the interval $[0, 1]$, respectively. In our scheme, we assign the basic probability value with the direct observation trust value. For example, if mobile user 1 believes that mobile user B is trustworthy, then the basic probability value for each possible hypothesis is:

$$m_1(H) = T_{A1}^D,$$
$$m_1(\overline{H}) = 0,$$
$$m_1(U) = 1 - T_{A1}^D, \tag{4.17}$$

where T_{A1}^D represents the trust value of mobile user 1 from the direct observation of mobile user B.

Oppositely, if mobile user 1 thinks that user B is untrustworthy, the basic probability value will be:

$$m_1(H) = 0,$$
$$m_1(\overline{H}) = T_{A1}^D,$$
$$m_1(U) = 1 - T_{A1}^D. \tag{4.18}$$

Concerning the formal definition of the belief function, let Ω be the universe of discourse, i.e., the set representing all possible states of the considered system.

$\Omega = \{trustworthy, untrustworthy\}$. The power set 2^{Ω} refers to the set of all subsets of Ω, here $2^{\Omega} = \{\varnothing, \{trustworthy\}, \{untrustworthy\}, \Omega\}$. Any hypothesis A_i refers to a subset of 2^{Ω}, and is mapped into a basic probability $m(A_i)$ that reflects the proportion of total belief assigned to hypothesis A_i. These two conditions should be satisfied: $m(\varnothing) = 0$ and $\sum_{A_i \subseteq \Omega} m(A_i) = 1$. For any hypothesis B, the belief function is defined as

$$beli(B) = \sum_{A_i \subseteq B} m(A_i), \tag{4.19}$$

which represents the strength of the evidence that supports hypothesis B's provability.

Dempster's Rule of Combining Belief Functions

Based on the definition of belief function, the Dempster-Shafer theory combines multiple users' belief. Assuming that $beli_1(B)$ and $beli_2(B)$ are two belief functions over the same universe of discourse, Ω. the orthogonal sum of $beli_1(B)$ and $beli_2(B)$ can be defined as

$$beli(B) = beli_1(B) \oplus beli_2(B)$$
$$= \frac{\sum_{i,j,A_i \cap A_j = B} m_1(A_i) m_2(A_j)}{\sum_{i,j,A_i \cap A_j \neq \varnothing} m_1(A_i) m_2(A_j)}, \tag{4.20}$$

where $A_i, A_j \subseteq \Omega$.

In our scenario, the combined degree of belief from mobile user 1 and mobile user 2 can be calculated as follows [37]:

$$m_1(H) \oplus m_2(H) = \frac{1}{K}[m_1(H)m_2(H) + m_1(H)m_2(U)$$
$$+ m_1(U)m_2(H)],$$

$$m_1(\overline{H}) \oplus m_2(\overline{H}) = \frac{1}{K}[m_1(\overline{H})m_2(\overline{H}) + m_1(\overline{H})m_2(U) \tag{4.21}$$
$$+ m_1(U)m_2(\overline{H})],$$

$$m_1(U) \oplus m_2(U) = \frac{1}{K}m_1(U)m_2(U),$$

where

$$K = m_1(H)m_2(H) + m_1(H)m_2(U) + m_1(U)m_2(U)$$
$$+ m_1(U)m_2(H) + m_1(U)m_2(\overline{H}) + m_1(\overline{H})m_2(\overline{H}) \tag{4.22}$$
$$+ m_1(\overline{H})m_2(U).$$

Following the rule of combination of belief, we can combine more degree of belief from other mobile users. Based on the Dempster-Shafer theory, T_{AB}^N is defined as:

$$Tr_{AB}^{InD} = m_1(H) \oplus m_2(H) \ldots \oplus m_n(H), \qquad (4.23)$$

where there are totally between mobile user A and mobile user B.

4.4 Problem Formulation

In this section, we formulate an optimization problem of the integrated trust-based social network with the 3C framework with MEC, caching, and D2D communications. We assume that a mobile user requests for a video content to the integrated network. For the network operator, it should decide which BS or a D2D transmitter is assigned to serve the requesting user, whether or not the video transcoding (i.e., MEC) should be performed, and whether or not newly emerged contents should be cached. The network operator need a comprehensive consideration of many factors, including: the wireless channel conditions, whether or not the requested content is stored at the local cache, whether or not the content version is matched up, the computational capacity, the trustworthiness of a D2D transmitter. Here, we consider dynamic scenarios, i.e., the available network conditions are varying with time. We exploit deep Q-learning algorithm to solve the formulated optimization problem.

In order to obtain the optimal policy, identifying the system's states, actions, and reward functions is necessarily required, which will be described in more details below.

4.4.1 System State

The system states for a subscriber s_l requesting video v_i at time slot $t \in \{0, 1, \ldots, T-1\}$ mainly includes five components: the realization $\Upsilon_{s_l}^{c_p}(t)$ of the random variables $\gamma_{s_l}^{c_p}$ (channel state), the realization $F_{s_l}^{c_p}(t)$ of the random variables $f_{s_l}^{c_p}$ (computation capability), the realization $X_{v_i}^{c_p}(t)$ of the random variables $x_{v_i}^{c_p}$ (content indicator), the version indicator $y_{v_i}^{c_p}$, and the trust value index $Tr_{s_l}^{c_p}$ for all the video providers, i.e., $\forall p \in \mathscr{C}$. Consequently, the system state vector can be described as follows.

$$\begin{aligned}
\chi_{s_l}(t) = \big[& \Upsilon_{s_l}^{c_1}(t), \Upsilon_{s_l}^{c_2}(t), \ldots, \Upsilon_{s_l}^{c_{K+M}}(t), \\
& F_{s_l}^{c_1}(t), F_{s_l}^{c_2}(t), \ldots, F_{s_l}^{c_{K+M}}(t), \\
& X_{v_i}^{c_1}(t), X_{v_i}^{c_2}(t), \ldots, X_{v_i}^{c_{K+M}}(t), \qquad (4.24) \\
& y_{v_i}(t), \\
& Tr_{s_l}^{c_1}(t), Tr_{s_l}^{c_2}(t), \ldots, Tr_{s_l}^{c_{K+M}}(t) \big].
\end{aligned}$$

Here, the trust index for all the BSs should be set to 1, i.e., $Tr_{s_l}^{c_p}(t) = 1, \forall c_p \in \mathscr{C}_K$; meanwhile, this index for all the D2D transmitters should be less than 1, i.e., $Tr_{s_l}^{c_p}(t) < 1, \forall c_p \in \mathscr{C}_M$. $y_{v_i}(t)$ denotes the indicator for whether or not the cached video version matches with the requested version. If yes $y_{v_i}(t) = 1$, if no $y_{v_i}(t) = 0$.

4.4.2   System Action

In our system, the network controller decides which video provider is assigned to the subscribed user, whether or not the computation offloading (video transcoding) should be performed, and whether or not the video provider should cache the new video. The network controller's composite action for subscriber s_l is given by

$$\mathbf{a}_{s_l}(t) = \{\mathbf{a}_{s_l}^{comm}(t), \mathbf{a}_{s_l}^{comp}(t), \mathbf{a}_{s_l}^{cache}(t)\}, \tag{4.25}$$

where the row vectors $\mathbf{a}_{s_l}^{comm}(t)$, $\mathbf{a}_{s_l}^{comp}(t)$, $\mathbf{a}_{s_l}^{cache}(t)$ are interpreted in more details as follows:

- The first row vector $\mathbf{a}_{s_l}^{comm}(t)$ is defined as

$$\mathbf{a}_{s_l}^{comm}(t) = [a_{s_l,c_1}^{comm}(t), a_{s_l,c_2}^{comm}(t), \ldots, a_{s_l,c_{K+M}}^{comm}(t)], \tag{4.26}$$

where $a_{s_l,c_p}^{comm}(t)$ is the association indicator, and $a_{s_l,c_p}^{comm}(t) \in \{0, 1\}$. If the subscribed user is associated with video provider c_p, $a_{s_l,c_p}^{comm}(t) = 1$, otherwise $a_{s_l,c_p}^{comm}(t) = 0$.

- The second row vector $\mathbf{a}_{s_l}^{comp}(t)$ is defined as

$$\mathbf{a}_{s_l}^{comp}(t) = [a_{s_l,c_1}^{comp}(t), a_{s_l,c_2}^{comp}(t), \ldots, a_{s_l,c_{K+M}}^{comp}(t)], \tag{4.27}$$

where $a_{s_l,c_p}^{comp}(t)$ indicates the computation offloading decision, and $a_{s_l,c_p}^{comp}(t) \in \{0, 1\}$. If the network controller has decided that the computation task should be performed at the provider c_p's device, $a_{s_l,c_p}^{comp}(t) = 1$; on the other hand, if the computation is decided to be executed on subscriber's own mobile device, $a_{s_l,c_p}^{comp}(t) = 0$.

- The third row vector $\mathbf{a}_{s_l}^{cache}(t)$ is defined as

$$\mathbf{a}_{s_l}^{cache}(t) = [a_{s_l,c_1}^{cache}(t), a_{s_l,c_2}^{cache}(t), \ldots, a_{s_l,c_{K+M}}^{cache}(t)], \tag{4.28}$$

where $a_{s_l,c_p}^{cache}(t)$ means the whether or not the video provider c_p should cache the newly-emerged video or the new version, and $a_{s_l,c_p}^{cache}(t) \in \{0, 1\}$. If the network controlled decides that c_p caches the video, $a_{s_l,c_p}^{cache}(t) = 1$, otherwise $a_{s_l,c_p}^{cache}(t) = 0$.

4.4.3 Reward Function

In this chapter, we set the system reward to be the total revenue (i.e., utility) of the network operator. The network operator charges the subscribed user s_l for associating with the video provider, which is denoted as λ_{s_l} unit price per bps. On the condition that video transcoding (computation offloading) is decided to be executed on video provider's side, the operator can charge for its computing service, which is defined as ν_{s_l} unit price per bps. In addition, if the network controller decides to let the BSs or D2D transmitters cache the new video or new version, the operator gets a potential revenue on estimated backhaul save, which is denoted as κ_{s_l} unit price per Hz.

On the other hand, the operator has to pay for the rented spectrum and backhaul bandwidth. The unit price for the usage of spectrum is defined as δ_{c_p} per Hz for provider c_p. Moreover, if video transcoding is performed, a certain amount of energy will be consumed for the computing, the network operator is obliged to pay η_{c_p} unit price for per consumed Joule. The cost of caching the video content in the memory of provider c_p is denoted as ς_{c_p} unit price per unit space.

The system reward for serving subscribed user s_l requesting video v_i is defined as the network operator's total revenue, and it is formulated as a function of the system states, and actions. The system actions determine whether or not the reward can be optimized. Here, we define the reward function for a specific subscribed user s_l as:

$$
\begin{aligned}
R_{s_l}^{v_i}(t) &= \sum_{c_p \in \mathscr{C}} \left[R_{s_l,c_p}^{\text{comm}}(t) + R_{s_l,c_p}^{\text{comp}}(t) + R_{s_l,c_p}^{\text{cache}}(t) \right] \\
&= \sum_{c_p \in \mathscr{C}} a_{s_l,c_p}^{\text{comm}}(t) \left(\lambda_{s_l} \, Tr_{s_l}^{c_p}(t) \Upsilon_{s_l}^{c_p}(t)(1 - w^{\text{comp}} a_{s_l,c_p}^{\text{comp}}(t)) - \delta_{c_p} B \right. \\
&\quad \left. - (1 - X_{v_i}^{c_p}(t)) \sigma^{c_p} \Upsilon_{s_l}^{c_p}(t) \right) \\
&\quad + \sum_{c_p \in \mathscr{C}} (1 - y_{v_i}^{c_p}) a_{s_l,c_p}^{\text{comp}}(t) (\nu_{s_l} Tr_{s_l}^{c_p}(t) \frac{F_{s_l}^{c_p}(t) o_{s_l}^{v_i}}{q_{s_l}^{v_i}} - \eta_{c_p} q_{s_l}^{c_p} e_{c_p}) \\
&\quad + \sum_{c_p \in \mathscr{C}} a_{s_l,c_p}^{\text{cache}}(t) (\kappa_{s_l} Tr_{s_l}^{c_p}(t) \Upsilon_{s_l}^{c_p}(t) - \varsigma_{c_p} o_{s_l}^{v_i}).
\end{aligned}
\tag{4.29}
$$

The above reward function is composed of three terms, i.e., the earnings from communications, computing and caching, respectively. Here, the trust index $Tr_{s_l}^{c_p}(t)$ is added to each incoming revenue term. For the first communication earnings, $\lambda_{s_l} Tr_{s_l}^{c_p}(t) \Upsilon_{s_l}^{c_p}(t)(1 - w^{\text{comp}} a_{s_l,c_p}^{\text{comp}}(t))$ denotes the available earnings for providing the video transmitting service to subscriber s_l. Note that if video transcoding is decided to be performed, the video will be compressed and the corresponding income will be reduced. $\delta_{c_p} B$ is the cost for consuming the radio spectrum bandwidth. $(1 - X_{v_i}^{c_p}(t)) \sigma^{c_p} \Upsilon_{s_l}^{c_p}(t)$ is the cost for possible consumed backhaul bandwidth: if $X_{v_i}^{c_p}(t) = 1$, i.e., the requested video exists in the cache of provider c_p, no backhual cost is needed; otherwise $X_{v_i}^{c_p}(t) = 0$, the video has to be fetched from the Internet and backhaul cost is unavoidable. Note that σ^{c_p} is a very important parameter. It is not realistic to build up a D2D communication when the requested video does not exist in its cache. Thus, σ^{c_p} should be set to be an extremely large number for c_p, $\forall c_p \in \mathscr{C}_M$. For the second computing earnings, $(1 - y_{v_i}^{c_p})$ is used to indicate whether or not the version is matching. $v_{s_l} Tr_{s_l}(t)^{c_p} \dfrac{F_{s_l}^{c_p}(t)o_{s_l}^{v_i}}{q_{s_l}^{v_i}}$ represents the gaining for implementing video transcoding, and $\eta_{c_p} q_{s_l}^{c_p} e_{c_p}$ is the expenditure for the energy consumption, where e_{c_p} denotes consumed energy for running one CPU cycle. For the third caching earnings, $\kappa_{s_l} Tr_{s_l}(t)^{c_p} \Upsilon_{s_l}^{c_p}(t)$ is the estimated future income of the saved backhaul bandwidth if the new video or new version is decided to be cached. Here the saved backhaul bandwidth is equal to the wireless communication rate. At last, $\varsigma_{c_p} o_{s_l}^{v_i}$ denotes the cost for caching the content.

$R_{s_l}^{v_i}(t)$ is the system's immediate reward, i.e., the network operator gets $R_{s_l}^{v_i}(t)$ at state $\chi_{s_l}(t)$ when action $\mathbf{a}_{s_l}(t)$ is performed in time slot t. However, a maximum immediate value is not a guarantee for the maximum long-term future rewards. Thus, we should also consider a big picture. A future reward with a discount factor φ is reasonable, which can be denoted as

$$R_{s_l, v_i}^f = \max_{\pi} \mathbb{E}\left[\sum_{t=0}^{t=T-1} \varphi^t R_{s_l}^{v_i}(t)\right], \qquad (4.30)$$

where π represents a Q-learning policy, i.e., a series of actions where a specific action will be executed given a specific system state, and φ^t approximates to be zero when T is large enough. In fact, a condition for terminating the training process should be set.

The objective of adopting deep Q-learning into our network model is to help find an optimization policy that can maximize the cumulated future rewards for the network operator.

4.5 Simulation Results and Discussions

In this section, we evaluate the performance of proposed scheme using computer simulations. We use TensorFlow [41] in our simulations to implement deep Q-learning. For performance comparison, the following four algorithms are presented: (1). Proposed scheme, which considers MEC, caching, D2D, as well as both direct observation and indirect observation; (2). Proposed scheme w.o. indirect observation, which does not consider indirect observation; (3). An existing scheme w.o. mobile edge computing [42]; and (4). An existing scheme w.o. D2D communications [2].

4.5.1 Simulation Settings

In the simulations, we consider an MSN consisting of 5 BSs, and 15 D2D transmitters. The radius of the cell is set to 500 m. Because D2D communications typically perform within short ranges, we use a clustered-based distribution model [43], where multiple users are located within one cluster with a radius of 50 m. In the network, there are 20 subchannels, each of which has a bandwidth of 180 KHz. The transmit powers of D2D transmitter and BS are 24 and 46 dBm, respectively. The noise spectral density is -174 dBm/Hz. A loss model $35.3 + 37.6 \log(d(m))$ [44] is used. In addition, block fading with a block size of 100 is assumed for channel fading. Moreover, we assume that there are totally 10 types of contents distributed in the network, and each content cache state follows the Markov model. We set the cache state transition probability of staying in the same state in the BS as 0.6 (0.3 in the D2D transmitter), and set the transition probability from one state to another as 0.4 (0.7 in the D2D transmitter). We further assume a Zipf popularity distribution in the simulations, with $\theta = 1.5$ [45]. The computation states of MEC servers follow the Markov model. We assume that the computation state of the MEC server can be very low, low, medium, high, and very high.

We assume that there are two types of D2D transmitters in the network: normal nodes, which share the content and perform computing normally, and compromised nodes, which modify contents maliciously. The BSs are assumed to be not compromised due to the physical security of BSs. We also assume that the number of compromised nodes is much less than the total number of nodes in the network. The attackers are independent. Hence, there is no collusion attack in the network.

In our simulations, we used a GPU-based server, which has 4 Nvidia GPUs with version GTX TITAN. The CPU is Intel Xeno E5-2683 v3 with 128G memory. The software environment we utilized is TensorFlow 0.12.1 with Python 2.7 on Ubuntu 14.04 LTS.

The training process of the deep convolutional networks are shown in Fig. 4.3. We formulate the 5 BSs and 15 D2D transmitters as the rows, and the 5 levels of computational capacity as the columns in a grid image. Whether or not the required

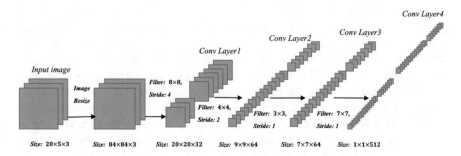

Fig. 4.3 The training process of the deep convolution networks

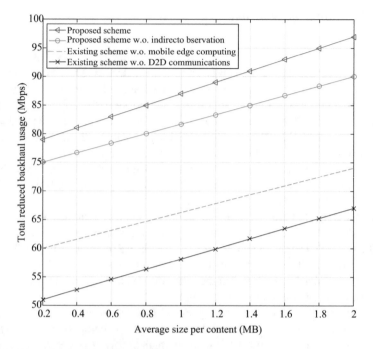

Fig. 4.4 The total reduced backhaul usage v.s. average size per content

content is in the local cache is characterized by different colors in each small grid. The initial input image of size $20 \times 5 \times 3$ is firstly resized into $84 \times 84 \times 3$. Through 4 layers of convolutional operations, the output is 512 nodes, and these nodes will be fully connected to be trained in the deep reinforcement learning.

4.5.2   Simulation Results

The effects of average size per content on the total reduced backhaul usage is shown in Fig. 4.4. There two malicious D2D transmitters, and there are 4 types of contents in the system. We can see from Fig. 4.4 that the total reduced backhaul usage increases with the increase of the average size per content. This is because a larger content size will increase the fee for caching the content, which induces a lower gain of caching utility, and thus the system is reluctant to cache the contents in the network. Compared to the existing schemes without mobile edge computing and D2D communications, the proposed scheme has larger total reduced backhaul usage due to the benefits of mobile edge computing and D2D communications. In addition, without indirect observation, the accuracy of trust evaluation is lower in the proposed scheme, which induces a lower gain of mobile edge computing and caching, and lower total reduced backhaul usage.

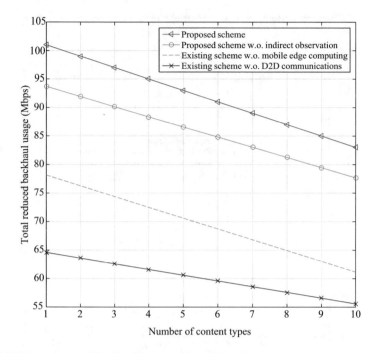

Fig. 4.5 The total reduced backhaul usage v.s. the number of content types

Figures 4.5 and 4.6 show the effects of the number of content types on the total reduced backhaul usage and the total utility, respectively. In these simulations, the average size per content is 1.8 MB, and there are two malicious D2D transmitters. We can see from both figures that the reduced backhaul usage and the total utility decrease as the number of content types increases. This is because more content types in the system will lead to the scenarios that a requesting mobile user is unlikely to find the specific content from the BSs and D2D transmitters due to the limited storage. In addition, more content types will result in reduced the popularity of all contents according to the Zipf popularity distribution, thus decreasing caching gain.

Figure 4.7 shows the effects of the required numbers of CPU cycles for video transcoding. From Fig. 4.7, we can see that the required number of CPU cycles has no effects on the total utility of the existing scheme without mobile edge computing. This is because the total utility will not be changed with different required numbers of CPU cycles if mobile edge computing is not available in the system. In the proposed scheme and the existing scheme without D2D communications, the total utility decreases exponentially with the increase of the required number of CPU cycles for video transcoding. This is because a larger number of required number of CPU cycles will result in a higher consumed computation energy, and consequently a lower gain of computation utility. Therefore, the total utility decreases with the required number of CPU cycles for video transcoding.

Next, we study the performance of the proposed social trust scheme. Assume that, at episode 14K, D2D transmitter 1 changes its maliciousness to 0.6. Our

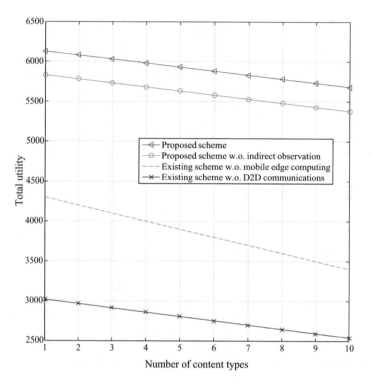

Fig. 4.6 The total utility v.s. the number of content types

goal is to observe the accuracy of trust tracking in this scenario where the malicious behavior of a D2D transmitter changes rapidly. Figure 4.8 shows the trust tracking of the system using direct observation with Bayesian inference and the proposed scheme using both direct observation with Bayesian inference and indirect observation with the Dempster-Shafer theory. We can observe from Fig. 4.8 that only direct observation can result in inaccurate trust values. By contrast, the proposed scheme can track the trust value accurately with both direct observation and indirect observation.

The number of malicious D2D transmitters in the network also has a significant impact on the performance the network. Here, We investigate the system utility with the number malicious D2D transmitters, from 1 to 5, in a 15 D2D transmitter environment. The basic parameter is the same as above. Figure 4.9 shows that, as the number of malicious D2D transmitters increases, the utility drops dramatically. When the number of malicious D2D transmitters reaches to one third of the total number of D2D transmitters in the network, the utility decreases to about half of the utility in the network with 1 malicious nodes. From this figure, we can see that the network is deeply affected by the number of malicious D2D transmitters.

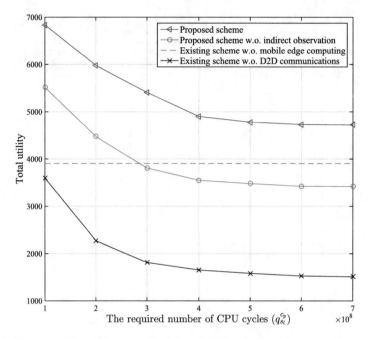

Fig. 4.7 The total utility v.s. the required number of CPU cycles for video transcoding

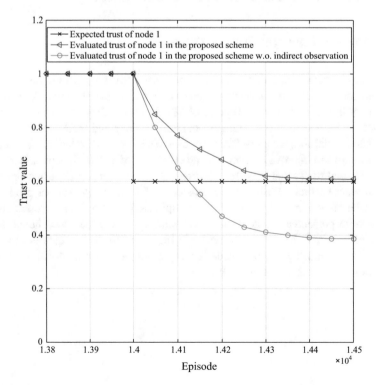

Fig. 4.8 The trust value v.s. the episodes

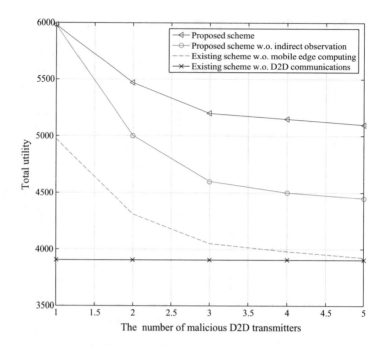

Fig. 4.9 The total utility v.s. the number of malicious D2D transmitters

4.6   Conclusions and Future Work

MSNs have become one of the most important networking paradigms in future wireless mobile networks. Recent advances of wireless mobile networks can have significant impacts on the performance of MSNs. In this chapter, we studied recent advances in mobile edge computing, in-network caching and D2D communications in MSNs. In addition, we considered the knowledge of social relationships in these new paradigms to improve the security and efficiency of MSNs. Specifically, we presented a social trust scheme with both direct observation using Bayesian inference and indirect observation using the Dempster-Shafer theory. We further proposed a deep Q-learning approach to study this complicated system. Extensive simulation results were presented to show the effectiveness of the proposed scheme. Since energy efficiency is becoming increasingly important in the big data era [46, 47] Future work is in progress to consider energy-efficient resource allocation strategy in the proposed integrated framework.

References

1. N. Vastardis and K. Yang, "Mobile social networks: Architectures, social properties, and key research challenges," *IEEE Comm. Surveys Tutorials*, vol. 15, no. 3, pp. 1355–1371, Third Quarter 2013.
2. Z. Su and Q. Xu, "Content distribution over content centric mobile social networks in 5G," *IEEE Comm. Mag.*, vol. 53, no. 6, pp. 66–72, Jun. 2015.
3. ETSI, "Mobile-edge computing – introductory technical white paper," *ETSI White Paper*, Sep. 2014.
4. N. Kumar, S. Zeadally, and J. J. P. C. Rodrigues, "Vehicular delay-tolerant networks for smart grid data management using mobile edge computing," *IEEE Comm. Magazine*, vol. 54, no. 10, pp. 60–66, Oct. 2016.
5. C. Wang, C. Liang, F. R. Yu, Q. Chen, and L. Tang, "Computation offloading and resource allocation in wireless cellular networks with mobile edge computing," *IEEE Trans. Wireless Comm.*, vol. 16, no. 8, pp. 4924–4938, Aug. 2017.
6. Y. Meng, C. Jiang, H. H. Chen, and Y. Ren, "Cooperative device-to-device communications: Social networking perspectives," *IEEE Network*, vol. 31, no. 3, pp. 38–44, May 2017.
7. K. Wang, F. R. Yu, and H. Li, "Information-centric virtualized cellular networks with device-to-device (D2D) communications," *IEEE Trans. Veh. Tech.*, vol. 65, no. 11, pp. 9319–9329, Nov. 2016.
8. Y. Zhang, E. Pan, L. Song, W. Saad, Z. Dawy, and Z. Han, "Social network aware device-to-device communication in wireless networks," *IEEE Trans. Wireless Comm.*, vol. 14, no. 1, pp. 177–190, Jan. 2015.
9. P. Hui, J. Crowcroft, and E. Yoneki, "Bubble rap: Social-based forwarding in delay-tolerant networks," *IEEE Trans. Mobile Comput.*, vol. 10, no. 11, pp. 1576–1589, Nov. 2011.
10. E. Bulut and B. K. Szymanski, "Friendship based routing in delay tolerant mobile social networks," in *Proc. IEEE GLOBECOM*, Miami, FL, USA, Dec. 2010, pp. 1–5.
11. S. Wang, X. Zhang, Y. Zhang, L. Wang, J. Yang, and W. Wang, "A survey on mobile edge networks: Convergence of computing, caching and communications," *IEEE Access*, vol. 5, pp. 6757–6779, Mar. 2017.
12. K. Zhang, Y. Mao, S. Leng, Q. Zhao, L. Li, X. Peng, L. Pan, S. Maharjan, and Y. Zhang, "Energy-efficient offloading for mobile edge computing in 5G heterogeneous networks," *IEEE Access*, vol. 4, pp. 5896–5907, Aug. 2016.
13. J. Liu, Y. Mao, J. Zhang, and K. B. Letaief, "Delay-optimal computation task scheduling for mobile-edge computing systems," in *Proc. IEEE Int. Symp. Inf. Theory (ISIT)*, Barcelona, Spain, Jul. 2016, pp. 1451–1455.
14. S. Sardellitti, G. Scutari, and S. Barbarossa, "Joint optimization of radio and computational resources for multicell mobile-edge computing," *IEEE Trans. Signal Inf. Process. over Netw.*, vol. 1, no. 2, pp. 89–103, Jan. 2015.
15. X. Chen, L. Jiao, W. Li, and X. Fu, "Efficient multi-user computation offloading for mobile-edge cloud computing," *IEEE/ACM Trans. Networking*, vol. 24, no. 5, pp. 2795–2808, 2016.
16. X. Chen, J. Wu, Y. Cai, H. Zhang, and T. Chen, "Energy-efficiency oriented traffic offloading in wireless networks: A brief survey and a learning approach for heterogeneous cellular networks," *IEEE J. Sel. Areas Commun.*, vol. 33, no. 4, pp. 627–640, Jan. 2015.
17. J. Gu, W. Wang, A. Huang, and H. Shan, "Proactive storage at caching-enable base stations in cellular networks," in *Proc. IEEE Int. Symp. on Personal Indoor and Mobile Radio Comm. (PIMRC)*, London, UK, Sept. 2013, pp. 1543–1547.
18. B. Bai, L. Wang, Z. Han, W. Chen, and T. Svensson, "Caching based socially-aware D2D communications in wireless content delivery networks: a hypergraph framework," *IEEE Wireless Commun.*, vol. 23, no. 4, pp. 74–81, Aug. 2016.
19. A. Sengupta, S. Amuru, R. Tandon, R. M. Buehrer, and T. C. Clancy, "Learning distributed caching strategies in small cell networks," in *Proc. 11th Int. Symp. Wireless Commun. Syst. (ISWCS)*, Aug. 2014, pp. 917–921.

20. Y. Guan, Y. Xiao, H. Feng, C.-C. Shen, and L. J. Cimini, "Mobicacher: Mobility-aware content caching in small-cell networks," in *Proc. IEEE GLOBECOM*, Austin, TX, USA, Feb. 2014, pp. 4537–4542.
21. Y. He, C. Liang, F. R. Yu, N. Zhao, and H. Yin, "Optimization of cache-enabled opportunistic interference alignment wireless networks: A big data deep reinforcement learning approach," in *Proc. IEEE ICC'17*, Paris, France, May 2017, pp. 1–6.
22. Y. He, F. R. Yu, N. Zhao, V. Leung, and H. Yin, "Software-defined networks with mobile edge computing and caching for smart cities: A big data deep reinforcement learning approach," *IEEE Commun. Magazine*, Sept. 2017.
23. Y. Zhou, F. R. Yu, J. Chen, and Y. Kuo, "Resource allocation for information-centric virtualized heterogeneous networks with in-network caching and mobile edge computing," *IEEE Trans. Veh. Tech.*, vol. 66, no. 12, pp. 11 339–11 351, Dec. 2017.
24. C. Wang, C. Liang, F. R. Yu, Q. Chen, and L. Tang, "Computation offloading and resource allocation in wireless cellular networks with mobile edge computing," *IEEE Trans. Wireless Comm.*, vol. 16, no. 8, pp. 4924–4938, Aug. 2017.
25. Y. He, N. Zhao, and H. Yin, "Integrated networking, caching, and computing for connected vehicles: A deep reinforcement learning approach," *IEEE Trans. Veh. Tech.*, vol. 67, no. 1, pp. 44–55, Jan. 2018.
26. Y. He, F. R. Yu, N. Zhao, V. C. Leung, and H. Yin, "Software-defined networks with mobile edge computing and caching for smart cities: A big data deep reinforcement learning approach," *IEEE Commun. Magazine*, vol. 55, no. 12, pp. 31–37, Dec. 2017.
27. Z. Su and Q. Xu, "Security-aware resource allocation for mobile social big data: A matching-coalitional game solution," *IEEE Trans. on Big Data*, vol. PP, no. 99, pp. 1–1, 2017.
28. X. Hou, Y. Li, M. Chen, D. Wu, D. Jin, and S. Chen, "Vehicular fog computing: A viewpoint of vehicles as the infrastructures," *IEEE Trans. Veh. Tech.*, vol. 65, no. 6, pp. 3860–3873, Jun. 2016.
29. A. Asadi, Q. Wang, and V. Mancuso, "A survey on device-to-device communication in cellular networks," *IEEE Commun. Surveys Tutorials*, vol. 16, no. 4, pp. 1801–1819, Fourth Quarter 2014.
30. C. Yang, J. Li, P. Semasinghe, E. Hossain, S. M. Perlaza, and Z. Han, "Distributed interference and energy-aware power control for ultra-dense D2D networks: A mean field game," *IEEE Trans. Wireless Comm.*, vol. 16, no. 2, pp. 1205–1217, Feb. 2017.
31. K. Hamedani, L. Liu, R. Atat, J. Wu, and Y. Yi, "Reservoir computing meets smart grids: attack detection using delayed feedback networks," *IEEE Trans. Ind. Inf.*, vol. 14, no. 2, pp. 734–743, Feb. 2018.
32. R. Atat, L. Liu, H. Chen, J. Wu, H. Li, and Y. Yi, "Enabling cyber-physical communication in 5G cellular networks: challenges, spatial spectrum sensing, and cyber-security," *IET Cyber-Phys. Syst.Theory Appl.*, vol. 2, no. 1, pp. 49–54, Apr. 2017.
33. I. R. Chen, F. Bao, and J. Guo, "Trust-based service management for social Internet of things systems," *IEEE Trans. Dependable and Secure Computing*, vol. 13, no. 6, pp. 684–696, Nov. 2016.
34. F. Parzysz, M. Di Renzo, and C. Verikoukis, "Power-availability-aware cell association for energy-harvesting small-cell base stations," *IEEE Trans. Wireless Commun.*, vol. 16, no. 4, pp. 2409–2422, Apr. 2017.
35. H. S. Wang and N. Moayeri, "Finite-state Markov channel-a useful model for radio communication channels," *IEEE Trans. Veh. Tech.*, vol. 44, no. 1, pp. 163–171, Feb. 1995.
36. H. Gomaa, G. G. Messier, C. Williamson, and R. Davies, "Estimating instantaneous cache hit ratio using markov chain analysis," *IEEE/ACM Trans. Netw.*, vol. 21, no. 5, pp. 1472–1483, Oct. 2013.
37. T. M. Chen and V. Venkataramanan, "Dempster-Shafer theory for intrusion detection in ad hoc networks," *IEEE Internet Comput.*, vol. 9, no. 6, pp. 35–41, Nov. 2005.
38. R. Changiz, H. Halabian, F. R. Yu, I. Lambadaris, H. Tang, and C. M. Peter, "Trust establishment in cooperative wireless networks," in *Proc. IEEE MilCom'10*, San Jose, CA, Nov. 2010, pp. 1074–1079.

39. C. Zouridaki, B. L. Mark, M. Hejmo, and R. K. Thomas, "Hermes: A quantitative trust establishment framework for reliable data packet delivery in MANETs," *Journal of Computer Security*, vol. 15, no. 1, pp. 3–38, Jan. 2007.
40. S. Bu, F. R. Yu, P. Liu, P. Manson, and H. Tang, "Distributed combined authentication and intrusion detection with data fusion in high-security mobile ad hoc networks," *IEEE Trans. Veh. Tech.*, vol. 60, no. 3, pp. 1025–1036, Mar. 2011.
41. M. Abadi, A. Agarwal *et al.*, "Tensorflow: Large-scale machine learning on heterogeneous systems," *arXiv:1603.04467*, Nov. 2015.
42. Y. Zhao, W. Song, and Z. Han, "Social-aware data dissemination via device-to-device communications: Fusing social and mobile networks with incentive constraints," *IEEE Trans. Services Computing*, vol. PP, no. 99, pp. 1–1, 2017.
43. B. Kaufman and B. Aazhang, "Cellular networks with an overlaid device to device network," in *Proc. Asilomar Conf. Signals, Systems & Computers (ACSSC)*, Pacific Grove, CA, Oct. 2008, pp. 1537–1541.
44. D. H. Lee, K. W. Choi, W. S. Jeon, and D. G. Jeong, "Two-stage semi-distributed resource management for device-to-device communication in cellular networks," *IEEE Trans. Wireless Commun.*, vol. 13, no. 4, pp. 1908–1920, Apr. 2014.
45. P. Blasco and D. Gunduz, "Learning-based optimization of cache content in a small cell base station," in *Proc. IEEE Int. Conf. Commun. (ICC)*, Sydney, Australia, Jun. 2014, pp. 1897–1903.
46. J. Wu, S. Guo, J. Li, and D. Zeng, "Big data meet green challenges: big data toward green applications," *IEEE Systems Journal*, vol. 10, no. 3, pp. 888–900, Sept. 2016.
47. J. Wu, S. Guo, J. Li, and D. Zeng, "Big data meet green challenges: Greening big data," *IEEE Systems Journal*, vol. 10, no. 3, pp. 873–887, Sept. 2016.

Printed in the United States
By Bookmasters